Migrant Communication Enterprises

LANGUAGE, MOBILITY AND INSTITUTIONS

Series Editors: **Celia Roberts**, *King's College London, UK* and **Melissa Moyer**, *Universitat Autònoma de Barcelona, Spain*

This series focuses on language and new ways of looking at the challenges facing institutions as a result of the mobility and connectedness characteristic of present day society. The relevant settings and practices encompass multilingualism, bilingualism and varieties of the majority language and discourse used in institutional settings. The series takes a wide-ranging view of mobility and also adopts a broad understanding of institutions that incorporates less studied sites as well as the social processes connected to issues of power, control and authority in established institutions.

Full details of all the books in this series and of all our other publications can be found on http://www.multilingual-matters.com, or by writing to Multilingual Matters, St Nicholas House, 31–34 High Street, Bristol BS1 2AW, UK.

LANGUAGE, MOBILITY AND INSTITUTIONS: 3

Migrant Communication Enterprises

Regimentation and Resistance

Maria Sabaté i Dalmau

MULTILINGUAL MATTERS
Bristol • Buffalo • Toronto

Library of Congress Cataloging in Publication Data
A catalog record for this book is available from the Library of Congress.
Sabaté i Dalmau, Maria, 1981–
Migrant Communication Enterprises: Regimentation and Resistance/
Maria Sabaté i Dalmau.
Language, Mobility and Institutions: 3
Includes bibliographical references and index.
1. Intercultural communication–Barcelona. 2. Migrant labor–Language–Barcelona.
3. Communication in small groups–Barcelona. 4. Telecommunication–Barcelona.
I. Title.
P94.65.B37S23 2014
306.44'094672–dc23 2014009372

British Library Cataloguing in Publication Data
A catalogue entry for this book is available from the British Library.

ISBN-13: 978-1-78309-218-5 (hbk)
ISBN-13: 978-1-78309-217-8 (pbk)

Multilingual Matters
UK: St Nicholas House, 31–34 High Street, Bristol BS1 2AW, UK.
USA: UTP, 2250 Military Road, Tonawanda, NY 14150, USA.
Canada: UTP, 5201 Dufferin Street, North York, Ontario M3H 5T8, Canada.

Website: www.multilingual-matters.com
Twitter: Multi_Ling_Mat
Facebook: https://www.facebook.com/multilingualmatters
Blog: www.channelviewpublications.wordpress.com

Copyright © 2014 Maria Sabaté i Dalmau.

All rights reserved. No part of this work may be reproduced in any form or by any means without permission in writing from the publisher.

The policy of Multilingual Matters/Channel View Publications is to use papers that are natural, renewable and recyclable products, made from wood grown in sustainable forests. In the manufacturing process of our books, and to further support our policy, preference is given to printers that have FSC and PEFC Chain of Custody certification. The FSC and/or PEFC logos will appear on those books where full certification has been granted to the printer concerned.

Typeset by R. J. Footring Ltd, Derby
Printed and bound in Great Britain by the CPI Group (UK Ltd), Croydon, CR0 4YY

To the Nobodies

The Nobodies
The nobodies: nobody's children, owners of nothing. The nobodies: the no ones, the nobodied, running like rabbits, dying through life, screwed every which way.

Who are not, but could be.
Who don't speak languages, but dialects.
Who don't have religions, but superstitions.
Who don't create art, but handicrafts.
Who don't have culture, but folklore.
Who are not human beings, but human resources.
Who do not have faces, but arms.
Who do not have names, but numbers.
Who do not appear in the history of the world, but in the police blotter of
   the local paper.

Eduardo Galeano, 1993 (excerpt from *El Libro de los Abrazos*, p. 52; translation by Cedric Belfrage)

Los Nadies
Los nadies: los hijos de nadie, los dueños de nada. Los nadies: los ningunos, los ningüneados, corriendo la liebre, muriendo la vida, jodidos, rejodidos.

Que no son, aunque sean.
Que no hablan idiomas, sino dialectos.
Que no profesan religiones, sino supersticiones.
Que no hacen arte, sino artesanías.
Que no practican cultura, sino folklore.
Que no son seres humanos, sino recursos humanos.
Que no tienen cara, sino brazos.
Que no tienen nombre, sino número.
Que no figuran en la historia universal, sino en la crónica roja de la prensa local.

Contents

Figures ix
Tables xi
Acronyms xii
Transcription Conventions xiv
Acknowledgements xvii

1 New Steps in the Sociolinguistics of Globalisation: The Critical Exploration of Migrant Institutions of Resistance in Late Capitalism 1
    The Context: Technology-Empowered Migrant Populations in Urban Catalonia 8
    Method and Data: The Network Ethnographic Window 16
    Overview of the Chapters 25

2 The Rise of Anti-Migrant Governmentality: Prelude to the Emergence of *Locutorios* 29
    The Post-Social State: Exclusionary Dataveillance Systems and Covert Linguistic Regimes 30
    The Telecommunications Sector: Global Capitalistic Dynamics and Ineffective Commercial Multilingualism 34

3 *Locutorios* as Challengers to Established Political-Economic Orders and Sociolinguistic Regimes 59
    From Autochthonous Local Businesses to Alternative Institutions of Transnational Survival 60
    Locutorios as Successful Transnational Points of 'Meetingness' and Mundane Resistance Practices 70

4 The Self-Provision of Technological Capital in *Locutorios*:
   A Diversity of ICT-Mediated Networking Practices   84
       Individual Mobility Projects and Subversive Communication
           Technology Tactics   85
       Transnational Family Units and the Collectivisation of ICT   93
       Maintaining Emotional Ties: *Doing* Family from a Distance   98

5 *Locutorio* Voices: Language and Literacy in Migrant-Regulated
   Discursive Spaces   107
       The Organisation of Silenced Multilingualisms in a Spanish-
           Unified Floor   110
       The 'Everyone's Spanish' Paradox: Subversion and Self-
           Discipline in Prevailing Linguistic Regimes   138

6 *Locutorios* as Migrant Spaces of 'Mismeeting' and Conflictive
   Togetherness   148
       Migrant Identities and Power Dynamics in Non-Mainstream
           Worlds   150
       Fighting Linguistic Exploitation: The Language and Identity
           Resources of the Abused   160

By Way of Conclusion: Informal Migrant Shelters in Which to
Critically Explore the Mundane Alphabets of the Future   171

Notes   176
References   181
Index   203

Figures

1.1 Percentage of community (Spanish nationals and foreign residents) using ICT (mobile phones, computers and internet) in Catalonia, 2010 — 9
1.2 Administrative regions in Catalonia — 17
2.1 The state's citizenship regimes and legal barriers. 'Compulsory registration plan for users of prepaid phone cards. Identify yourself!' — 32
2.2 Image from the campaign against the 'digital exclusion' of migrants, *Plan Avanza* — 33
2.3 The telecommunications sector's unrealistic management of linguistic diversity. The MundiMóvil packaging presents the phrase 'The world in your hands' in 10 different languages — 54
3.1 Numbers of *locutorios* registered in Vallès Occidental, January 2001 and March 2009 — 64
3.2 Multinationals' competition techniques against 'ethnic' *locutorio* businesses. The *'Locutori mòbil'* (Orange) and the *'mini-locutorio'* (Movistar) — 68
3.3 The migrants' self-distribution of key resources (information). Leaflet offering legal services with free consultation in the *locutorio* of El Paso — 78
3.4 Services provision in non-recognised migrant languages. Leaflet with a Movistar discount plan in Urdu, translated from Spanish on a migrant's initiative — 79
3.5 The intercultural mediation practices of *locutorio* workers. Scan of a receipt of a successful money transfer written in Spanish — 81
4.1 The market's discursive and visual representations of ICT-mediated 'transnational communication' — 86

4.2 The market's construction of economically based transnational family relationships. Telefónica's campaign 98
5.1 Tokenistic commercial multilingualism in *locutorios*. Pictures of the front and the back of a Habibi call card 111
5.2 Non-elite written multilingual practices in *locutorios*. Shabbir's notebook 111
5.3 Bilingual written practices in Catalan and Spanish. Shabbir's notebook 134
5.4 The symbolic place of non-elite allochthonous codes in *locutorios*. Unconventional 'no smoking' sign handwritten in Arabic and Spanish 136
5.5 Hidden-in-public allochthonous codes in *locutorios* 138
5.6 The social uses and meanings of 'everyone's Spanish'. Room-for-rent advertisements posted by the same user 143
5.7 'Everyone's Spanish' in migrant-tailored corporate information. Discount plans by Siempre Latina 145
5.8 Self-disciplining practices in written Standard Spanish. Two room-for-rent advertisements for the same flat 147
6.1 The migrants' social forms of differentiation in language and in identity display. Room-for-rent advertisements posted on the walls of the *locutorio* in El Paso 151
6.2 Technological support for *locutorio* employees. (a) Cabin telephone counter and (b) *locutorio* computer programs 169

Tables

2.1 Migrant-oriented mobile phone operators in the Spanish state (2009) — 37
2.2 The linguistic landscape of the telecommunications sector in Barcelona, 2007–09: Customer service call centres — 47
3.1 Percentage of foreign population and number of *locutorios* by town, Vallès Occidental, 2001 — 65
3.2 Percentage of foreign population and number of *locutorios* by town, Vallès Occidental, 2008 — 66
3.3 Average net income generated at the *locutorio* of El Paso (2007–09) — 67
4.1 Migrants' use of *locutorio* telephone cabins — 87
6.1 Constructions of identity in discourse: Terms of address for a *locutorio* worker — 163

Acronyms

ACPI	Asociación para el Conocimiento de la Población Inmigrante (Association for Information about Immigrant Populations)
ALYCE	Asociación de Locutorios y Cíbers Españoles (Spanish Association of *Locutorio* and Cybercafés)
BT	British Telecommunications plc
CCTV	closed-circuit television
CEO	chief executive officer
CHILDES	Child Language Data Exchange System
CIEN	Research group Comunicació Intercultural i Estratègies de Negociació (UAB)
CIF	Certificado de Identificación Fiscal (corporate identification number, for tax purposes)
CLA	Research group Cercle de Lingüística Aplicada (UdL)
CMT	Comisión del Mercado de las Telecomunicaciones (Spanish Telecommunications Market Commission)
CTNE	Compañía Telefónica Nacional de España (National Telecommunications Company of Spain)
CV	curriculum vitae
DNI	Documento Nacional de Identidad (identification card for Spanish nationals)
EU	European Union
FOBSIC	Fundació Observatori per la Societat de la Informació a Catalunya (Observatory for the Information Society of Catalonia Foundation)
GDP	gross domestic product
ICT	Information and communication technology
ID	identification document

Idescat	Institut d'Estadística de Catalunya (Official Statistics Institute of Catalonia)
INE	Instituto Nacional de Estadística (Spanish National Statistics Institute)
ITU	International Telecommunication Union
LIDES	Language Interaction Data Exchange System
MMS	Multimedia Messaging System
MVNO	mobile virtual network operator
NGO	non-governmental organisation
NIE	identification number for foreign residents living in Spain
ONTSI	Observatorio Nacional de las Telecomunicaciones y de la Sociedad de la Información (Spanish Observatory of Telecommunications and the Information Society)
PC	personal computer
SETSI	Secretary of State for Telecommunications and the Information Society
SIM	Subscriber Identification Module
SMS	Short Message Service
STSI	Secretariat for Telecommunications and the Information Society in Catalonia
UAB	Universitat Autònoma de Barcelona
UdL	Universitat de Lleida
UNESCO	United Nations Educational, Scientific and Cultural Organization
USB	universal serial bus (standard computer port)
VAT	value added tax

Transcription Conventions

The excerpts taken from the oral interview data that I present in this book have been transcribed following the conventions of the Language Interaction Data Exchange System for transcribing and analysing multilingual language practices, the LIDES (LIPPS Group, 2000), which is based on MacWhinney's (2000 [1991]) international transcription standard for language acquisition data, the CHILDES.

In order to meet the aims of this piece of research, though, two LIDES transcription procedures have been adapted as follows. Firstly, I have used different type styles (plain, bold, underline and italic) instead of language tags in order to make the transcriptions as readable and amenable as possible, following Codó's transcription procedures (2008: xi–xiii). Secondly, I have employed no intra-turn rising (-¿) or intra-turn exclamation (-!) contours. Instead, I have marked intra-sentential questions [¿] and exclamations [!] by making use of the scopes (< >), as in CHILDES. I was not fully satisfied with these two LIDES contours because it seemed that they could just have made up for, or simply replaced, conventional punctuation marks, undermining many features of the non-standard, hybrid talk that I aimed to analyse (see also Pavlenko, 2007).

What follows are the detailed transcription conventions that I have used in this book:

Dependent Tiers

%act: participants' actions while speaking
%add: addressee(s) of a particular turn
%com: situational information and comments about the main tier
%tra: free translation into English of the main tier

Language Coding (typefaces)

Plain: Spanish
Italics: Catalan
<u>Underlined</u>: English
Bold: undecidable language

Transcription Conventions

+^	quick uptake or latching
+...	trailing off
xxx	unintelligible material
www	untranscribed confidential information
[...]	omitted exchange
#	pause
[=! text]	paralinguistics, prosodics (e.g. laughter)
[>]	overlap follows
[<]	overlap precedes
(¿)	best guess
< >	scope
[!]	stressing
:	lengthened vowel
::	longer lengthened vowel
[/]	repetition
[//]	retracing, reformulation
+"/.	quotation provided in the next line
+"	quotation follows in this line

Intonation Contours

.	end-of-turn falling contour
¿	end-of-turn rising contour
!	end-of-turn exclamation contour
-,.	end-of-turn fall–rise contour
-.	intra-turn falling contour
-,	intra-turn fall–rise contour
,,	tag question

Each data excerpt starts with two pieces of contextual information, presented under two headers taken from LIDES. The first header (@ Location) provides information on the date and the geographical location of the interviews (the *locutorio*, the bars or the street). Needless to say, for confidentiality reasons I do not provide any particular details by means of which the informants, the cities, the institutions or the spaces that I mention could be identified.

The second header (@Bck) provides the background information on the interaction. It includes a brief explanation of the topics of the exchange, and it also details the participants in a generic speaker code that uses *RES for the researcher and the three main letters for the participants, referring to either their professions or to their fictitious names (for example, I use *CON for consultant and *ABD for informant Abdelouahed).

All the exchanges have been reproduced verbatim in their various non-standard forms of talk. I provide a free translation for the turns in languages other than English in a dependent tier (%tra) introduced below the main tier. I have made every effort to capture and reproduce messy, broken language practices in the translations, including non-Western expressions voiced in Spanish, clarified through the analysis and the discussion.

As this volume is part of a socially engaged, critical sociolinguistic ethnography, I finish this section with the reminder that the transcription choices we make have tremendous implications for the data analysis because they are always affected by our research goals and interests and, consequently, reflect our particular positioning and perceptions (Bucholtz, 2000; Codó *et al.*, 2012; De Fina & Georgakopoulou, 2008; Heller, 2006 [1999], 2007a, 2010a; Ochs, 1979; Sarangi & Candlin, 2001). Thus, the transcription procedures used here should be read as an integral part of the research objectives that I had in mind as an active participant in the construction of the knowledge about the migrants' resistance practices in *locutorios* which I aim to present in the following chapters.

Acknowledgements

Many, many people deserve to be acknowledged and thanked for their contributions to this project. To start with, without the help of the persons who shared their silenced lives as undocumented migrants in the outskirts of Barcelona I simply would not have been able to write this book. It was they who opened the doors of our town to an extent that I would never have imagined. To all the informants who participated in this ethnography, particularly to the Pakistani community of the neighbourhood that I here call El Paso, my most heartfelt thanks. A special word of gratitude goes to Naeem, the best *locutorio* worker in town, who coped with my presence and with my insistent questions for more than two years. *Hero puray Gujranwale da!*

I also owe much of what I have learnt from this experience to Monica Heller and Melissa G. Moyer, who have been a constant and impressive source of knowledge and encouragement. Many of the reflections that I present here were nested and developed within the research group CIEN (Intercultural Communication and Negotiation Strategies), based at the Universitat Autònoma de Barcelona (UAB), my first academic home, and I deeply thank each of its present and past members.

I am also very thankful to Alexandre Duchêne and to Joan Pujolar, as well as to David Atkinson, Iris Bachmann, Mike Baynham, Gabriele Budach, Eva Codó, Hortènsia Curell i Gotor, Ana Fernández Montraveta, Maria Rosa Garrido i Sardà, Michael Kennedy, Clare Mar-Molinero, Ulrike H. Meinhof, Sònia Oliver, Adriana Patiño, Maite Puigdevall, Celia Roberts, Adela Ros and Patrick Stevenson for having shared their bright ideas about *locutorio* matters, and for their tremendous help at different stages of this project.

I am grateful, too, to the Department of Anthropology's 2006–07 Masters cohort at the University of Toronto, for the best insightful discussion sessions, especially to Daniella Jofré, Aaron Kappeler and Mireille McLaughlin. Also, I benefited from the invaluable contributions of the research teams at the Centre for Transnational Studies in the Department of Modern Languages at the University of Southampton, on the one hand, and at the Institute of Multilingualism at the University and Haute Ecole Pédagogique of Fribourg, on the other – they both really felt like home.

I am really very pleased, too, with my new colleagues in the English and Linguistics Department at the Universitat de Lleida (UdL), where the last lines of this book were written, for their warm welcome and for their incredible support. Our meetings within the research group Cercle de Lingüística Aplicada (CLA) have also proved very fruitful!

Thanks to Neus Sabaté i Barrieras for her kind technical help with the maps; to Angelica Carlet for double-checking the phonetic symbols that have been used here; and to Marc Carreras, Najma Husain, Youssef Ihmadi, Safae Jabri, Nargus Karim, Gemma Puigvert, Daniel Ramon, Pimpila Thanaporn and Lin Xie, too, for their assistance with the data in Modern Standard Arabic, Mandarin Chinese, Greek, Hindi, Panjabi, Russian and Urdu.

I am also indebted to the staff of the 23 town councils of the municipalities of the Vallès Occidental area who, like the telecommunications agents that participated in this project (linguistic engineers, mobile phone operator entrepreneurs and call centre assistants), found a gap in their busy agendas to provide me with crucial information not only about the Spanish telecommunications sector but also about *locutorio* locations and regulations in Catalonia. Of course, the Spanish telecommunications companies, as well as the *locutorio* businesses that I visited, follow the ever-changing speedy dynamics of the globalised new economy. Consequently, I want to highlight that what I say about them here is just my understanding of a small piece of their history, which must already have changed in very substantial ways from the time I finished my fieldwork, in 2009.

I want to acknowledge, too, the outstanding work of the Multilingual Matters team, which consists of Angharad Bishop, Martisse Foster, Tommi Grover, Laura Longworth, Elinor Robertson, Anna Roderick and Sarah Williams, as well as Ralph Footring. I wish to thank, as well, the two editors of this book series, Celia Roberts and Melissa G. Moyer, for their patience and professionalism.

The protection of the informants' anonymity and the confidentiality of the data analysed in this book (including tape-recorded and photographed materials) have been ensured by the ethics committee at the Universitat Autònoma de Barcelona (the CEEAH; registration file 725H) as well as by Col·lectiu Ronda, a Catalan cooperative of lawyers committed to situations of exclusion, marginalisation and racism. Any shortcomings are of course mine. Likewise, I am solely responsible for all that is presented here, as well as for the ways in which it is presented.

This work has greatly benefited, too, from the predoctoral grants PIF 429-01-1/07 (UAB) and BE1 00362 2009 (AGAUR), and from the following research group grants: Multilingualism and Mobility: Linguistic Practices and the Construction of Identity (FFI2011-26964/FILO, MINECO); The Management of Multilingualism in Institutional Settings (HUM2007-61864/FILO, MINECO); The Management of the Linguistic Reception of Newcomers in a Public Institution and a Non-Governmental Institution (2007 ARAFI 00018, AGAUR); and 2009 SGR 1340 (AGAUR).

I dedicate this book to Elisabet, for the sweetest memories, and to all of you, great family and extremely encouraging friends from here and there and, of course, from that little town in Catalonia called Vallbona d'Anoia. Your commitment, understanding and support have accompanied me from the introduction to the conclusions. Many thanks for always, always, having been on board. *Moltes gràcies*!

1 New Steps in the Sociolinguistics of Globalisation: The Critical Exploration of Migrant Institutions of Resistance in Late Capitalism

> *By looking at how ICTs are constructed, accepted, adopted or adapted in migration contexts, we are putting forward a new 'lens' for analysing migration in the information society.*
> Adela Ros, Elisabet González, Antoni Marín and Papa Sow (2007: 33)

Well into the second decade of the 21st century, it is not new to state that the current life trajectories, family configurations and work prospects of transnational migrant populations are largely established, maintained and enhanced (although not readily propelled) by ICT (Castells, 2000 [1996]; Urry, 2007; Vertovec, 2001, 2007, 2010), in particular by the mobile phone – the *social glue* for global trotting (Vertovec, 2009: 54). Indeed, there are, globally, 86 mobile phones per 100 people (ITU, 2012a: 2) and, although technological disconnection has by no means disappeared (Donner, 2008a, 2008b; Edjabe & Pieterse, 2001; Hannam et al., 2006; Inda & Rosaldo, 2002; Sheller & Urry, 2006), internet use has become almost demotic. The extent, intensity and velocity of networked flows of contact and information continue to increase (Appadurai, 1996; Castells et al., 2007a; Goggin, 2006; Horst & Miller, 2006; Ling, 2004; Steinbock, 2005), and migrant populations, now engaged with forms of global capitalist consumption (Bauman, 2000, 2005; Ros & Boso, 2010; Vertovec, 2001), are at the cutting edge in the adoption of ICT (Lam & Rosario-Ramos, 2009; Lam & Warriner, 2012; Panagakos & Horst, 2006).

Thus, over the past few years, ICTs have become a powerful source of *social capital* for 21st-century transnational survival. They are now a pivotal way for mobile migrant populations to access more material and symbolic resources and to establish and maintain powerful social networks, in the not so new information and communication age (Alonso & Oiarzabal, 2010; Androutsopoulos, 2006a, 2006b, 2007; Castells, 2009; Danet & Herring, 2007; Peñaranda Cólera, 2011; Ros & Boso, 2010; Ros *et al.*, 2007).[1]

And yet, by focusing on the benefits of the technological revolution of the late 1990s for transnational citizens, we have neglected the structural barriers (that is, the institutional citizenship regimes and economic orders) imposed upon the members of migrant social networks, who, being at the bottom of the social ladder, experience all sorts of digital divides as technology 'have-nots' (Bertot, 2003) – despite living in societies that belong to the so-called information-rich nation-states.

Indeed, in the area that I investigated – Catalonia, an officially Catalan–Spanish bilingual region with its own quasi-autonomous government located in the north-eastern part of the Spanish state[2] – many migrants who try to gain access to ICT today find themselves facing a series of marginalising legal hindrances, serious economic constraints and, above all, exclusionary communicative and linguistic hurdles. These barriers constitute the *regimentation practices* governing the citizenry of the 21st century, particularly for controlling (un)documented migrants, via the close regulation of their conditions of access to their primary tool for successful transnational networking: information and communication technology.

These regimes[3] are imposed by two extremely powerful institutions, which exercise their governmentality practices on citizens in very complex, closely interrelated manners, in a post-national scenario where advanced liberal forms of rule akin to those of the free market seem to be replacing the older public welfare state (Castells, 2012; Fairclough, 2006; Inda, 2006; Pujolar, 2007a, 2007b). The first institution is the post-social nation-state and its new market-driven techniques of 'registration of non-citizens' (Collins, 2005: 243) via ICT. The second is the telecommunications sector, which is one of the biggest sellers of mass consumption services and products and one of the most important segments of the private business world; it also regiments transnational peoples, by regulating their consumption habits worldwide.

Despite all sorts of obstacles (or perhaps because of them), migrants mobilise their own transnational resources and find subversive ways to access ICT, at the margins of these two powerful technopolitical institutions. Intrigued by their capacity for resilience, my main objective in the research was to investigate where, why, how, in what languages and with what consequences particular transnational populations resist nation-state power in present-day Catalonia. More specifically, I aimed to understand how the very different groups of migrants who, undocumented, are largely pushed to social disconnection and stagnation, but nevertheless manage to articulate a series of mundane resistance practices against state-imposed and market-driven exclusion, through their own technological power and their multilingual capitals, overcoming exclusionary social orders.

I explore such resistance practices in an alternative institution of bottom-up subversion and empowerment whereby variegated migrant networks establish their own ICT businesses in order to overcome the precariousness that they encounter in hostile host societies.[4] This space is the *locutorio*, essentially a type of *ethnic* call shop that sells telephony, internet, fax and money transfer services which, gradually established in many Catalan urban areas at the turn of the 21st century, became a unique informal meeting point for all sorts of transnational migrant networks seeking transnational survival.

From a socially engaged, critical approach to the sociolinguistics of globalisation (see Blommaert, 2003, 2010; Coupland, 2003, 2010; Duchêne et al., 2013; Fairclough, 2006; Heller, 2006 [1999], 2010a; Pennycook, 2010, 2012; Piller, 2011), I argue that *locutorios* are an excellent research space where to examine various sociolinguistic phenomena related to these pauperised citizens' daily resistance to structural social inequality. More specifically, I show that *locutorios* allow for the analysis of the unparalleled emergence and consolidation of truly alternative institutions of transnationalism, whereby migrant populations can gain a certain degree of individual and collective social agency,[5] in connection to, but from the margins of, what we may call mainstream society. I claim that these ventures show how different migrant populations have subverted anti-migrant regimes and, largely successfully, relocalised and inserted themselves into economic and sociopolitical orders in their own self-regulated, but highly hierarchised, social spaces, in the unexplored worlds of cosmopolitan Catalonia.

By providing an ethnographically grounded, fine-grained analysis of various *locutorio* phenomena, I hope to advance our knowledge of how migrants' ordinary resistance techniques and subversive tactics emanate from, and get inscribed into, their particular *languages, mobilities* and *institutions* – the three anchoring themes of the series in which the present volume appears.

Regarding mobilities, I present new findings concerning what transnational living actually consists of, but from a migrant-centred, participatory perspective which allows for the reflexive problematisation of hegemonic or taken-for-granted conceptions of the migrants' movements and interconnections, as well as of their 'demobilisations' and their stages of 'mooring' (Hannam et al., 2006: 2). In trying to understand their 'here and there' life trajectories, I hope to make a contribution, too, to the investigation of the 'internal' social structuration processes[6] of these highly diverse groups of transnational populations, providing a picture of the workings of the power dynamics and of the in- and out-group distribution of transnational resources. This includes, crucially, a detailed picture of their *actual* ICT-mediated communicative practices, particularly in relation to family configurations and physically distant relationships, and, finally, of their work prospects in Catalonia, at a time when the globalised new economy was seriously troubled (Castells, 2012; Harvey, 2010) and the Spanish state was dealing with a sovereign debt crisis.

Secondly, regarding institutions, I claim that *locutorios* also allow for the problematisation of our assumptions regarding what counts as 'institutions' today, since they definitely do not sit well with traditional notions or well established classifications of institutional settings, particularly those which draw classic distinctions between private institutions (related to the marketplace), public institutions (related to governmental bodies) and non-governmental institutions (mostly NGOs and volunteer agencies). Although *locutorios* have to follow the rules of the private business sector – its market rationalities and its commercial objectives, like any other ICT business (mobile phone operators, fax and photocopy shops or money transfer agencies) they are not directly dependent on the Spanish telecommunications market, since they are regulated, in an intra-group manner, by transnational networks of migrant entrepreneurs. Thus, *locutorios* may sustain (and may be sustained by) informal managerial work forms and grey market practices, for many sell or hire legal goods and services outside the formal circuits of commerce and outside the official distribution channels.

Besides, *locutorios* provide many vital resources (like food, information on legalisation procedures, job offers, or even shelter) that would normally be attributed to the help or the donations of either the public administration of a welfare state (and its ancillary agencies) or else to pro-migrant non-governmental organisations or associativism initiatives, for instance. Nevertheless, this provision of precious resources in these spaces is carried out from the margins, without following any public social responsibility or any altruistic or altermondialist agenda. Instead, *locutorios* are an alternative public 'resource centre' (Peñaranda Cólera, 2005; Roca i Albert et al., 2009) self-regulated by and for diverse groups of migrants, according to their own largely under-researched social stratification rules. Therefore, they are neither fully private nor fully public institutions, for which they point to a phase of 'late capitalism' (Heller & Duchêne, 2012: 3) that speaks of a shift towards the informalisation of nation-state economies, propelling the birth of altogether different, 21st-century hybrid institutions of migration and transnational living.

Finally, concerning the third domain, language, I suggest that *locutorios* allow for the privileged 'insider' observation of the migrants' management and organisation of their own (and of other migrants') heteroglossic language practices and ever-changing linguistic ideologies,[7] when these social players mobilise their myriad multilingual literacy and numeracy capitals in local interactions taking place solely among themselves. In particular, I argue that the discursive spaces of *locutorios* let us carefully explore the workings of the migrants' non-elite (i.e. non-valued and silenced) multilingualisms,[8] not only when they present themselves as transnational navigators in their host societies but also, more revealingly, when they fight for a voice of their own from within their social networks.

These migrant-regulated language battles and these competitions of linguistic capitals, which, of course, constitute the migrants' multiple fluid identity construction projects, also bring us closer to the language-mediated establishment of highly complex translocal social relationships of affinity and support, on the one hand, as well as of unseen rivalry and hatred, on the other. Thus, my third contribution in this book consists of an in-depth investigation of the migrants' particular sociolinguistic hierarchies and of their unconventional identity dialectics. I analyse how their self- or other-ascribed social categorisation tactics are mediated in these empowering institutions, where they can take the 'communicative floor' (Goffman, 1981) and, in fact, start to appropriate and colonise the 'urban linguistic landscapes' (Jaworski & Thurlow, 2010: 13) of Catalan towns, and indeed global cities like Barcelona (Martín-Rojo, 2012: 289–290).

With these three research avenues in mind, I ultimately attempt to provide a comprehensive picture of what *locutorios* actually mean *for* these migrant populations. And I do so by critically questioning, reformulating and going beyond deep-seated trivialising assumptions or preconceptions concerning present-day technology-endowed migration movements in Europe. I provide a new account of how multilingual transnational citizens fight anti-migrant citizenship regimes and make sense of, and voice, who they feel they are and of what piece of the global map they feel they inhabit, in the context of the Catalan network society.

At this point, some clarification is required concerning the use of six terms frequently used in this text: *migrant, ethnic business, transnationalism, globalisation, resistance* and *locutorio*. I employ the term *migrant* (rather than 'immigrant') for two reasons. Firstly, many mobile citizens in the Spanish state experience a series of unfinished migration trajectories which are highly complex. These journeys may imply living in various nation-states simultaneously (or paying long and short visits to multiple places), beyond leaving a 'home' to become a naturalised citizen in a 'landing country' (Solé et al., 2007, 2008; Vertovec, 2001). I believe that the term *migrant* avoids dichotomous *e*migrant/*im*migrant distinctions and better gathers the idea of 21st-century demographic movements. Secondly, the term *immigrant* or *inmigrante*, respectively used in the Catalan and the Spanish languages, have been connoted in very negative ways, and have at times become a racialised exclusionary term of address. With the use of the term *migrant* I try to dissociate myself from these pejorative connotations and from more ethnocentric approaches to mobile citizenship.[9]

By *ethnic business* – a term coined by a group of American economic sociologists in the early 1970s[10] – I mean 'those commercial ventures run by persons of foreign origin [...] independently of the characteristics of the services and products commercialised and of the geographical distribution of such businesses'[11] (Parella Rubio, 2005: 258). Thus, they are small or medium-size enterprises owned and run by both low- and middle-class re-territorialised populations whose main objective is to attract transnational migrant clients, though they may be convenient, too, for tourists and for a local clientele who are attracted by the cheaper products and services (Feliu et al., 2012; Íñiguez-Rueda et al., 2012; Moreras, 2007). In the present context, 'ethnic' businesses include migrant-tailored shops such as Chinese bazaars, Moroccan halal butcher shops, Latino hairdressers and Pakistani *locutorios*, which today are part and parcel of the Catalan local economy. The term, as one may expect, is controversial, since it has frequently been used to stress the ethnocultural (cum racial) 'differences' or 'exceptionalities' of these commercial ventures, again from an ethnocentric perspective (Beltrán et al., 2007; Glick Schiller & Çağlar, 2013; Parella Rubio, 2005; Pecóud, 2000; Rath & Kloosterman, 2000). However, 'ethnic' is the most widely accepted adjective employed, in Catalonia and elsewhere, as a descriptive category (*not* as a social category) simply indicating the origin of the entrepreneurs and that of the targeted clientele as well (nonetheless, the term is generally presented in inverted commas to denote these reservations).[12]

Following Vertovec (2009: i), I use the term *transnationalism* to make reference to 'the multiple ties and interactions linking people or institutions across the borders of nation-states [...] within the study of globalization'. That is, I employ it to talk about the sorts of interconnections which simultaneously have an effect on tiny localities and on big nation-states and which transform the new 'here', the old 'there', and the altogether different 'here and there' of these social players. I use the term *globalisation* only when I want to focus on the scale and scope of global interdependencies, and when

I need to stress the increase in the quantity, rapidity and intensity of such linkages (Da Silva *et al.*, 2007; Inda & Rosaldo, 2002). Thus, I do not employ it to highlight the newness of the phenomena, for I do not see mobilities and flows (or immobilities and spatial stasis), *per se*, as new phenomena (see discussions in Coupland, 2010; Pennycook, 2012; Sheller, 2011).

I do not understand the migrants' *resistance* to be a well articulated revolutionary anti-system movement. Rather, I see it as a pragmatic series of self-protective mundane reactions against institutionalised forms of social order (Gal, 2001; Horst & Miller, 2006; Kohler Riessman, 2000) or, more specifically, against given exclusionary governmental practices which, on a daily basis, are imposed upon repressed 'non-citizens' by nation-state bodies and by the global markets. These migrants' subversive attempts to skirt the bureaucratic machinery of power are not about victories, but about small, partial achievements and modest bottom-up strategies mobilised by those 'have-nots' for whom to exist is to resist (Alabajos, 2011). Therefore, these resistance tactics, which speak of 'coping', of resilience and of endurance of inflicted social inequality, in a tacit 'rhizomatic' way (i.e. spreading horizontally and vertically across and within migrant individuals), have the transformative effect of making collective transnational living more bearable (Castells, 2008; Scott, 1989). My argument throughout the book is that *locutorio* phenomena actually epitomize many of these resistance strategies, in an unparalleled manner.

I use this term, *locutorio*, in Spanish, for two reasons. Firstly, throughout the book I claim that, though they bear similarities to other ICT businesses (regular phone shops, telecentres, internet centres and cyber cafés), *locutorios* provide an unprecedented degree of social agency to migrant populations, for which they have become truly transnational migrant institutions of struggle, resistance and survival, rather than just mere specialised local shops. Secondly, I avoid translations into English, and I keep the name of such places in Spanish (and not in Catalan) because I want to highlight the fact that, paradoxically, the migrants who socialise there end up reproducing local linguistic normativity regimes which foster the use of a unified floor in Spanish, in turn silencing their own multilingual voices (as well as Catalonia's co-official language, Catalan) and reproducing the linguistic marginalisation in which they are forced to organise their transnational lives in this part of the world. The use of the term *locutorio* in the following chapters will remind the reader of the complexities of this phenomenon, which encompasses many of the social contradictions migrant populations have to navigate today, in discursive spaces which speak of both the process and the product of the globalised new economy, in the particular context of the Catalan network society, which is detailed in the following section.

The Context: Technology-Empowered Migrant Populations in Urban Catalonia

I frame this study in the context of the 'diversification' of diversity (Vertovec, 2007: 1024) of present-day Western societies like Catalonia where there has been an intensification of many different 'categories of migrants' (Blommaert & Rampton, 2011: 1) whose heterogeneity can be attributed to the complex interplay of a range of social factors, including: people's nationality and place of birth, their legality and citizenship statuses, life experiences and expectations, transnational trajectories and work prospects, as well as religious affiliations, identity attributes, and, of course, language backgrounds and literacy repertoires (Blommaert, 2012; Normann Jørgensen & Juffermans, 2011; Vertovec, 2006).

I address diversity in Catalonia by providing a brief summary of the five local contexts that are most relevant to the present study (most of which also concern many other European societies, too): the technological, demographic, legal, linguistic and labour conditions in which the multiplicity of skilled and unskilled migrants that I observed on the outskirts of Barcelona found themselves at the time when the fieldwork was undertaken, between 2007 and 2009. These five contextual aspects (analysed below in the order in which they have just been mentioned) will more clearly situate this particular piece of research, and they will be my point of departure for anchoring and for investigating, later on, the migrants' resistance practices and subversive ICT-mediated communication tactics in *locutorio* businesses.

Technological context

The first aspect that I need to mention concerns the technology boom which affected all populations in Catalonia at the turn of the 21st century, when the Spanish state, which had already joined the influential International Telecommunication Union (the United Nations' biggest sociopolitical body within the telecommunications realm; see ITU, 2011), became an 'information-rich' country occupying a privileged position right after the United States, the United Kingdom, Germany, France, Canada, Italy, the Netherlands, Switzerland and Belgium (Barnett, 2001; Paetsch, 1993). Catalonia then experienced a transition to the *network society*, a prevalent way of organising social life which has, at its epicentre, the social media gizmos and gadgets of the new millennium (Castells, 2000 [1996]; Castells *et al.*, 2007b). Migrant populations were crucially affected by this, for they very soon engaged in what has come to be known as the migrant *mobile-isation* (Castells, 2012; Miyata *et al.*, 2005; Urry, 2003; Wellman,

2001): an unforeseen social mobilisation via the use of ICT which rapidly empowered these transnational citizens with increased collective social agency in their host societies (Castells *et al.*, 2004; Peñaranda Cólera, 2011; Plataforma per la Llengua i Consum Català, 2011).

The impact of this ICT revolution has been such that transnational migrants today show a connectivity rate which is even higher than that of Spanish nationals, mostly with regard to the use of mobile phones for SMS communication (texting) (Castells *et al.*, 2007b: 53; Ros & Boso, 2010: 146).

Figure 1.1 presents a summary of the most comprehensive study on ICT use carried out by the Observatory for the Information Society of Catalonia Foundation (Fundació Observatori per la Societat de la Informació a Catalunya, FOBSIC) together with the Official Statistics Institute of Catalonia (Institut d'Estadística de Catalunya, Idescat) drawing on data from 5,455,432 informants aged between 16 and 74, including a representative sample of migrant populations (FOBSIC & Idescat, 2010).[13] It shows that foreign residents use the mobile phone as their primary means of communication (96.2%), more than non-migrants living in Catalonia (92.8%). The technology magazine of the Spanish newspaper *El País* reported that the migrant groups who most used mobile phones were those born in Eastern Europe, followed by those who originally came from the Maghreb area and then those from Latin America (CiberP@ís, 2007). While the percentage of habitual computer users (that is, the percentage of people who use a PC at least once a week) was similar for non-migrants and migrants (94.5% and

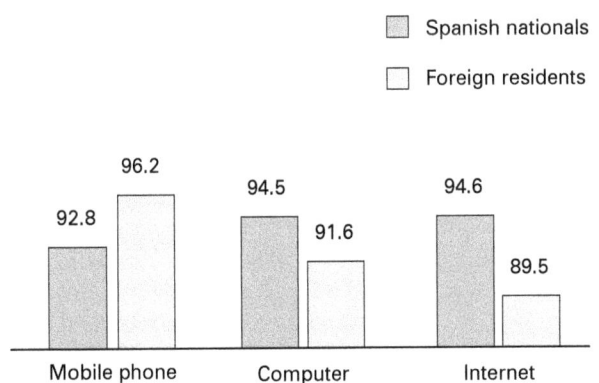

Figure 1.1 Percentage of community (Spanish nationals and foreign residents) using ICT (mobile phones, computers and internet) in Catalonia, 2010. *Source*: FOBSIC & Idescat (2010). Data assembled by author

91.6%, respectively), fewer foreign residents (89.5%) were habitual users of the internet than Spanish nationals, who used the internet as much as computers (with a percentage of internet use of 94.6%).

A closer look at age groups reveals that migrant adolescents and young adults are further *mobile-ised*, since those aged between 15 and 29 actually initiate *twice* as many calls, SMS and internet connections as non-migrant adolescents and young adults (Robledo, 2008). If we look at the type of internet uses, this same report (FOBSIC & Idescat, 2010: 28–29, 55) shows that migrants use the web to search for information concerning job vacancies and goods, services, travel and housing (90%), as well as for email (76.5%), chatting, instant messaging or blogging purposes (62.9%) and internet telephone calls (41.9%). Spanish nationals go online to search for information concerning goods and services, travel, housing and health issues (97.5%), as well as for email (91.5%), but use instant messaging to a much lesser extent (48.7%) than migrants do and make relatively few internet phone calls (16%).

The aim of presenting these data on the migrants' use of ICT is to show that these populations have become crucial transnational providers and consumers of digital *knowledge* and *information* (Alarcón & Garzón, 2011a; Diminescu, 2008; Katz, 2008; Ros & Boso, 2010). As the core members of their networks, their circulation of information and knowledge makes it easier for relatives, partners, friends and acquaintances to plan their migration projects; in particular, to plan mobility strategies and to take informed decisions regarding the protection of both the family unit and the family income (Peñaranda Cólera, 2011; Ros *et al.*, 2007; Vertovec, 2009).

Demographic context

The second contextual piece of information required for an understanding of *locutorios* concerns a change in the migrants' mobility patterns over the past few years. Until recently, mass migration to Catalonia, contrary to what happened in, say, the UK or the USA, was a largely new phenomenon. Between 2000 and 2011, the percentage of foreign residents in the region rose from 2.9% to 15.7%, out of a total population of 7,535,251 at the end of that decade (Idescat, 2010a, 2011). In fact, Spain as a whole was until very recently placed among the 10 top countries in the world in terms of the highest increase in the number of migrants (Arango, 2009: 54).

However, the growth of migrant populations then stalled (Idescat, 2010b; Secretaria de la Immigració, 2010). While the renewal of work permits increased by 30% between 2009 and 2010 (these still corresponded to migrants who could make use of a past immigration law to renew temporary residence), over the same period the numbers of new family reunification petitions and of new temporary residence permits each decreased by more than 500,000, partly due to the economic recession (Solé, 2011). At the same time, 15,581 documented foreign residents (mostly single men aged between 15 and 29 of Ecuadorean, Bolivian, Peruvian and, to a lesser extent, Pakistani and Chinese origin) left Catalonia between 2010 and 2011, either to return to their place of origin or to move to yet another country (INE, 2011: 3).

Legal context

The demographic picture is closely linked to recent changes concerning migrants' legal status. In 2009, the estimated number of undocumented people in the Spanish state was claimed to be one of the lowest for 10 years (Ministerio del Interior, 2012). Although today most migrants are documented, many of them (particularly non-EU residents) can still not vote in municipal elections, travel freely in the EU, hold public office or keep their original nationality (Vigers & Mar-Molinero, 2009: 174–177). This is so because they were granted temporary or permanent residence after having undergone one of the three extraordinary regularisation processes undertaken in 2000, 2001 and 2005 by the Spanish government as a way to officially recognise and 'legalise' a mass of undocumented transnational citizens (see Kostova Karaboytcheva, 2006). Since that time, with the more restrictive European directive on common standards for the expulsion of undocumented third-country nationals,[14] and with the 2009 modification of the Spanish Immigration Law (Ley Orgánica de Extranjería 4/2000), those migrants who at present are denied entry and those resident but who are denied some type of legal citizenship are now expelled from Catalonia (sometimes after a stay at a detention centre) via forced return (repatriation) or readmission to a third country (Ministerio del Interior, 2012).

Thus *locutorios* emerged at a time when a diversity of young adult migrant populations constituted, for the first time, a relatively large, long-term, mostly documented group of highly connected citizens networking in and from Catalonia.

Linguistic context

These transnational social networks of people born in other parts of Europe (30.4% of migrants in Catalonia in 2011), Africa (26.8%) and South America (25.7%) (Idescat, 2011) have changed Catalonia's demographic face, and they have also obviously had a very notable impact on its complex bilingual sociolinguistic situation, challenging the public and private hegemonic spaces that the Spanish and the Catalan languages occupied and occupy in Catalonia (Corona et al., 2013; Marshall, 2006; Newman et al., 2013; Pujolar, 2010; Torres-Pla, 2012). Thus, the fourth contextual piece of information which will be highly relevant for the analysis of *locutorios* is that, over the past 20 years, foreign residents have introduced more than 300 allochthonous codes (most notably German, English, Arabic, Tamazight, French, Fulah, Hindi, Italian, Mandinka, Panjabi, Portuguese, Quechua, Romanian, Russian, Soninke, Tagalog, Ukrainian, Urdu, Wolof, Wu and Mandarin Chinese, according to Linguamón, 2010) to metropolises such as Barcelona, where they have definitely started to occupy and appropriate many of their urban enclaves and linguistic landscapes (see also Çağlar, 2001; Generalitat de Catalunya, 2007; Martín-Rojo & Molina, 2012). This makes Catalonia a unique setting for a sociolinguistic exploration of the recent challenges that linguistic diversity poses to the European bilingual societies of late capitalism.

In Catalonia, the Spanish and the Catalan languages have long coexisted in very complicated ways, in a contested linguistic panorama which was, and is, the site of heated debates over Catalan language minority rights, Spanish linguistic uniformity and Catalan–Spanish bilingual education policies, in themselves connected to broader issues of nationhood and the political autonomy of Catalonia within a centralist Spanish nation-state (for a more comprehensive historical overview see Alarcón & Garzón, 2011b; Garzón, 2012; Pujolar, 1995, 2001; Woolard, 1989, 2003, 2006). Because Spanish is the dominant official language of the entire nation-state, and a high-status pioneering global lingua franca too (Del Valle, 2006; Mar-Molinero, 2006; Trenchs-Parera & Newman, 2009), the vast majority of foreigners in Catalonia seem to choose to learn Spanish first (Alarcón & Garzón, 2011a: 140; Fabà Prats, 2012: 45; Pujolar, 2010: 231; Torres-Pla, 2012: 27). This has occurred to the detriment of Catalan, which has a long historical and political trajectory of prosecution, minorisation and subordination. It is not recognised as an official language by the EU and is officially regarded only as the *co*-official (i.e. not *the* official) language and the '*llengua pròpia*' or 'vernacular' code of Catalonia (Generalitat de Catalunya, 2006). It is losing out in both private and public use, as the percentages of people using the language have decreased notably over the last 10 years, despite the fact that the number of Catalan speakers has remained stable (Fabà Prats, 2012: 41; Torres-Pla, 2012: 27; Torrijos, 2013).

With the incorporation of migrant populations into the cultural, political and socioeconomic workings of Catalonia, the Spanish and the Catalan languages have acquired newer social meanings. Basically, the (non-)use of Catalan and/or Spanish by transnational migrants has manifestly relaxed the normative expectations and sociolinguistic behaviours encapsulated in the older Catalans–Castellans dichotomy. This local binarism rendered Catalan as an identity marker and as a national symbol for skilled local middle-class Catalan families, and conversely linked the use of Spanish to the 'Castilian' identity marker, mainly for the large numbers of working-class 'internal' migrants who came to Catalonia from southern Spain six decades ago (Frekko, 2013; Pujolar & Gonzàlez, 2013; Woolard, 2009; Woolard & Frekko, 2013).

As a response to this increased linguistic diversity, the intricate relationships between Catalan and Spanish in the institutional arena have necessarily started to include the management of multilingualism, as seen in the Catalan government's initiatives like the comprehensive 'Education Plans' (Departament d'Ensenyament, 2012) emanating from the National Agreement on Immigration (Generalitat de Catalunya, 2008), which promotes the social use of Catalan as a shared, cohesive language of a cosmopolitan Catalan bilingual society respecting migrants' multilingualisms. However, despite these inclusive, post-modern 'civic' policies, the relatively scarce (though exemplary) practical initiatives directly addressing the migrants' linguistic incorporation into their host societies are, in fact, what Patten and Kymlicka (2003: 5) call 'pragmatic accommodations', and do not amount to official linguistic rights for speakers of non-elite codes inhabiting Catalonia.

This may account for the only partial recognition and the largely unsuccessful institutional management of the diverse communicative frames and linguistic capitals of transnational populations. This is apparent not only in the public sector – notably in healthcare (Martín-Rojo, 2010; Moyer, 2011), schooling (Sabaté i Dalmau, 2009) and public administration (Codó, 2008) – but also in the third sector, notably in a variety of (religiously inspired) volunteer social service organisations (see Garrido, 2010; Pujolar, 2007b, 2009). Similarly, the private business sector in Catalonia generally uses Spanish to welcome migrant populations (Pujolar, 2007a) or, alternatively, Spanish and Catalan and/or in some elite European linguae francae such as English, French and, to a lesser extent, German or Portuguese (ELAN. Cat, 2006; Generalitat de Catalunya, 2009; Solé Camardons & Torrijos, 2011). In the particular case of the telecommunications world, social media technology is more often than not offered to migrants in Spanish (this is particularly so for the former monopoly provider, the multinational Telefónica – see Canyelles, 2011: 142). Sometimes migrant ICT clients are catered for in English as an 'emergency code' for managing communication with multilingual customers, though this by no means happens systematically (Barcelona Activa, 2011: 12). Finally, only very marginally are these populations in Catalonia helped in non-elite allochthonous codes (usually either Modern Standard Arabic or Romanian). Consequently, this segment of the Spanish market leaves those who have no capitals in the Spanish (or English) language largely underserved and in need of an alternative provider: basically, the more multilingual, migrant-oriented *locutorios* (Sabaté i Dalmau, 2012a).

Labour conditions

The last contextual aspect of the successful consolidation of *locutorios* concerns the migrants' working conditions and work prospects. At the time of the fieldwork, the unemployment rate in Catalonia, which used to be the most industrialised urban area and which used to have the highest GDP within Spain (Solé, 2011: ix), reached 22.82% (INE, 2012). Many labour migrants who, as a strategy to access Europe, moved to Catalonia for economic reasons now face insecurity at the workplace. Most were recruited as unskilled workers regardless of their professional and educational backgrounds and now tend to occupy marginal and unwanted local employment niches (Moyer & Martín-Rojo, 2007; Solé & Alarcón, 2001), on temporary contracts (Elias Boada, 2011) and low salaries – the average annual gross salary for documented foreign residents is €17,881.75 (Observatori d'Empresa i Ocupació, 2012), approximately 51–61% less than that of Spanish residents (Carrasco Carpio & García Serrano, 2012: 15).

Their situation has recently become even more precarious: the unemployment rate among foreign residents in Catalonia increased steadily during the time of the fieldwork, and it went on to reach 39% for the first time in 2012 (Observatori d'Empresa i Ocupació, 2012: 4), one of the highest rates in Europe (Arango, 2009: 58), mostly affecting young Moroccan and Romanian men (Generalitat de Catalunya, 2011a: 3), 20% of whom have no access to any unemployment benefit or social aid (Migracat, 2011a). Partly as a consequence of these factors, it is estimated that approximately half a million migrants have started to engage in informal economic activities or 'grey market' practices (Carrasco Carpio & García Serrano, 2012: 15).

At the same time, though, there are other transnational groups of migrant entrepreneurs who, by contrast, have achieved labour autonomy and may enjoy some economic stability or improvement (and even upward social mobility), challenging hegemonic workplace regimes in Catalonia. As a reaction to the structurally blocked opportunities that they encounter in the unstable local labour market, some transnational migrants who were already skilled businesspersons before migrating have managed to set up their own 'ethnic' businesses (Arjona Garrido & Checa Olmos, 2006; Beltrán *et al.*, 2007; Cebrián de Miguel & Bodega Fernández, 2002), the *locutorios* being a good example (Peñaranda Cólera, 2005; Serra del Pozo *et al.*, 2003; Solé *et al.*, 2007). These are migrants who have accumulated capital, experience and several years of temporary residence as employees, but then registered with the social security office as self-employed and employers, instead of employees (Oso Casas & Villares Varda, 2005; Parella Rubio, 2005). Although in 2010 it started to decrease, by the end of fieldwork (in 2009) the number of self-employed[15] foreign residents had reached 93,179, which amounted to 23.42% of transnational migrants registered with the social security office in Catalonia (Observatorio Permanente de la Inmigración, 2010). They were mostly Chinese, Romanians, Bulgarians and Pakistanis, many working in the information and communication sectors, and in domestic services and in catering, with the demise of the construction sector (Migracat, 2011b; Secretaria de la Immigració, 2009).

The migrants belonging to this 'new breed of entrepreneurship on the make' (Castells & Portes, 1989: 11) now themselves formally or informally hire and profit from the cheap foreign labour of the globe's peripheries. While some (particularly Chinese and Pakistani) migrant businesspeople tend to employ persons from their extended transnational family units, in the Catalan network society this is now often extended to other members of the same ethnolinguistic communities with whom they hold strictly entrepreneurial and profit-motivated relations. This means that they now substitute family relations for relationships of production established with distant acquaintances with whom they have no emotional involvement (Castells, 1999; Parella & Cavalcanti, 2008).

The pool of transnational migrant workers that these entrepreneurs have readily available at present (in Catalonia as elsewhere) are the unprotected migrant employees looking for a job. These workers silently submit to exploitative work conditions due to the socioeconomic and legal status in which they are trapped. For the owners of *locutorios* they possess two unique 'added values' to make their bosses' 'ethnic' businesses succeed: the 'migrant baggage' and the 'insider' tacit knowledge of the consumption habits and distinct technology needs of the prospective (migrant) clientele; and the communicative capitals and wide range of non-elite languages – the unique multilingual and intercultural expertise – by means of which these workers can attract migrant technology consumers. Both values are forms of human capital that come at almost zero cost for these employers, for they are normally treated as mere uncompensated 'ethnic attributes', not as crucial payable workplace resources (Alarcón & Heyman, 2013).

In the particular context of the Catalan network society that I have summarised in this section, what is interesting is that these *locutorio* workers play a pivotal role in the unfolding of resistance practices in alternative institutions of transnationalism. This is basically because, by tirelessly providing technoliteracy capital (i.e. technology and literacy skills)[16] to non-Spanish-dominant or to non-literate multilingual clients, they individually help to articulate the connection of the disconnected. Therefore, they actually do the multilingual language job that the formal telecommunications market has left unaccomplished. It is for this reason that these *locutorio* workers became the privileged point of departure through which I gradually got to know about subversive migrant transnational life in the realms of the *locutorios*, as explained in detail in the following section.

Method and Data: The Network Ethnographic Window

This piece of research started in a *locutorio* when I began to investigate the intricate workings of migrant social networks in an urban area near Barcelona. I had been led there in tracking down the migrant populations' impressive everyday 'coping' with social stagnation and also their subversive ICT-mediated communication tactics against all sorts of adversities. I was immediately captivated, too, by the complex relationships established among competing or allied migrants groups, particularly by their largely unknown 'internal' fights concerning the (non-)legitimacy of particular non-elite linguistic repertoires, as well as by their habitation of fluid translocal migrant identities, which seemed difficult to grasp outside that migrant-governed space. It is precisely because of the uniqueness of these institutions and of the people who inhabited them that I find it necessary to detail some of the basic ethnographic steps that allowed me to undertake this project, which will also later allow for a better understanding of some of my claims.

Thanks to a Pakistani *locutorio* worker who kindly accepted my presence and, in the end, welcomed me inside his workplace, I conducted a two-year comprehensive ethnography whereby I could follow, observe and gradually connect the mobility trajectories, family configurations, work experiences, multilingual practices and ICT-mediated communication routines of a very heterogeneous group of 20 transnational migrants who in different ways all faced 'uncertainty about the future of one's livelihood' (Bauman, 2001: 41). They were all organised in and around that particular *locutorio*, which was located in a very marginal neighbourhood that I here call El Paso. This was a migrant-populated, impoverished working-class area with a high crime rate and frequent police raids. It was located on the periphery of a medium-sized

Figure 1.2 Administrative regions in Catalonia. The county of Vallès Occidental where fieldwork was conducted is indicated (darker shading). *Source*: Generalitat de Catalunya (2011b)

town on the outskirts of the metropolitan area of Barcelona, in the Catalan county of Vallès Occidental (Figure 1.2).

In order to understand the sociolinguistic regimes and social orders that regulated the *locutorio* under study, it is relevant to state that this particular county was characterised by the fact that 33.69% of its inhabitants were not born in Catalonia. At the time of the fieldwork, 12.09% of its population were transnational migrants, mostly born in South America and Africa (Idescat, 2011), and 21.60% were Spanish-speaking persons who came from other regions of the Spanish state during the 1950s and 1960s (Idescat, 2010c), when about 1.5 million Spanish nationals moved to Catalonia seeking economic betterment in the context of a precarious economy after the Spanish Civil War (Atkinson, 2000; Narotzky & Smith, 2006).

The county of Vallès Occidental had an unemployment rate of 17% in September 2011 (Copevo, 2011), although the three towns in which I did the fieldwork, particularly the one in which the *locutorio* that I observed was located, had rates which were substantially higher than that average (and than the average for Catalonia as a whole) during the same period.[17] In fact, the neighbourhood that I studied was publicly presented, in the maps provided by the local town hall, as an 'unemployment pocket' and as an 'ethnic enclave'.

The members of the multi-layered *locutorio* network that I investigated were a highly heterogeneous group of men and women aged between 27 and 52. Most were born in Pakistan, Morocco and Romania and some in Argentina, El Salvador, the Dominican Republic, Bolivia, Guinea, Equatorial Guinea, Cuba and Brazil. With the exception of a few who became *locutorio* workers or who still worked in the declining construction sector or in factories, the informants were largely unemployed and undocumented; part of their income derived from the local, very active, informal economy, regulated in an in-group manner. The vast majority had previously lived in other parts of Spain (some also in other parts of the EU, often the UK or Greece), and had settled in the neighbourhood of El Paso between 2002 and 2010. Their family configurations were also extremely diverse: single middle-aged men or young women who had migrated alone and were sharing a single room; extended families or various couples living together in sub-let apartments; or single families with children, living with Spanish individuals who needed the extra income from sharing their flat. Finally, they were also linguistically highly diverse, but navigated a local linguistic market where Spanish clearly figured highly over Catalan and, indeed, over any other language. Whereas some were non-schooled and non-literate in the Western communicative practices prevailing in Catalonia, others had a high command of both oral and written Spanish, as well as of English and Darija, for instance.

The type of mobile ethnography that I employed to get to the roots of this fluid web of migrants was *network ethnography* (Howard, 2002), also called *ethnography of networks* (Wittel, 2000). This is a 'transdisciplinary method [...] that should be especially useful for studying communication in modern organizations over new media' (Howard, 2002: 551). This type of ethnographic practice emerged from within those frameworks for social network analysis that are used to explore the links between given linguistic practices or behaviours and certain social orders regulating a particular place at a particular time. It emerged at the turn of the 21st century, when social network analysts seeking to understand 'worlds-in-the-making' (Sheller, 2011: 8) found it necessary to take mobility, distance, change and circulation as the point of departure for their investigations.

Apart from taking today's highly interconnected order of things into account, the network ethnographic method was also extremely useful in that it allowed for 'the examination of the nodes of a net and the examination of the connections and flows (money, objects, people, ideas, etc.) between these nodes' (Wittel, 2000: 5). Thus, it offered me the tools with which to gradually draw links between the intertwined and overlapping trajectories of the key components of a multifaceted transnational *locutorio* network and to interrogate and to operationalise the links between its 'nodes' (that is, the various individuals or families, their languages and key pieces of knowledge).

Moreover, precisely because it is based on what Marcus (1995) calls 'web-tracking strategies', this ethnographic method provided me with highly flexible research resources to move out from single, bounded or fixed 'sites' and to examine some broader 'fields' (Sinatti, 2008) with a remarkable degree of systematicity. This gave me a much wider perspective of the transnational network under study, and many more contextual insights into relevant *locutorio* phenomena. In practice, this means that I could observe the members of the *locutorio* network not only inside the *locutorio* (which I took as my point of departure) but also in many other discursive spaces, such as their (informal) workplaces and households, the two bars near the *locutorio*, the park and, above all, the street of our neighbourhood. Thus, I observed and drew connections between the members of the mobile *locutorio* network not only in each of these spaces but also *between*, *across* and *beyond* them. To a certain extent I defined the scope of the 'field' on-site, while conducting the actual research, by following the informants' paths, without conferring *a priori* schemes for such spatial allocations and mobility/immobility practices (Ardèvol *et al.*, 2008). This was crucial, as it allowed me to keep challenging my own presuppositions, (mis)conceptions and (mis)interpretations about their roles, relationships and activities in El Paso. All in all, it was actually this intriguing *locutorio* network (and its hierarchized re-distributions of material and symbolic resources, or crucial 'nodes') which, in the end, became my *research space*.

Of course, 'ethnographies on and in networks [...] might also be difficult to realize in a methodological sense' (Wittel, 2000: 23). Indeed, trying to understand a polycentric transnational migrant network by following it on the spot was definitely a challenge, for three main reasons. Firstly, I had many difficulties in limiting the scope of the research, and I often had to change, undo and try out several ethnographic paths. Even just deciding whom to follow around the *locutorio* was laborious (though extremely rewarding) and demanded long-term participant observation.

Secondly, letting the informants establish what route to follow forced me to take unexpected ethnographic roads that I could not plan for. This made it difficult for me to balance the 'working within' while avoiding being 'subject to' the informants (see Sarangi & Roberts, 1999, for a discussion on this matter). Nonetheless, we negotiated fruitful agreements and a 'nexus of collaborative [and non-collaborative] practice' (Jaffe, 2012: 334) concerning the degree of involvement in this project and the limits of our researcher/researched roles.

Thirdly, this particular network ethnography required constant personal involvement, which meant that I ended up sharing parts of my personal story to an extent that perhaps would not have been required with other qualitative methods of data collection.

What follows is a summary of the ethnographic steps that I took, as well as of the different types of data that I collected during the fieldwork, which will provide more specific information on how I applied the network ethnographic method to the transnational *locutorio* web (I explore this ethnographic practice in more detail in Sabaté i Dalmau, 2012b).

Unsurprisingly, gaining access to the *locutorio* network was the most difficult task. Already fascinated by the different *locutorios* that I, as a user, had frequently visited in different towns of Vallès Occidental, in 2007 I moved to a flat in El Paso (the area with the cheapest housing at that time), in a block that housed about 50 working-class people of very diverse origins (although mostly from southern Spain, Morocco and Peru). The popular *locutorio* that I investigated, one of the busiest in town, was conveniently located just downstairs.

In order to gain access to the *locutorio* network, I started checking my emails and printing my students' homework (in English) there daily, so that the *locutorio* worker as well as his clients could see that I was working at the nearby university (the UAB, 20 minutes from El Paso). I also got into the habit of sitting on the benches on the pavement in front of the *locutorio* every evening and engaged in informal conversations with as many clients as possible, letting myself be observed and interrogated about my job, my arrival in town, my marital status and the like. I also happened to meet clients of the *locutorio* regularly elsewhere in the neighbourhood, including the public transport stations, the outdoor cafés and the grocery stores.

My next step was to show the most frequent users of the *locutorio* my university identification card and to explain the research project to them, but when I asked for their help and participation in it, telling them exactly what I was planning to do, how and why, they were completely unimpressed, and I did not get very far with this first attempt. They were more impressed (after they had checked) by the fact that I was, indeed, a genuine neighbour. At that point they did give me informed consent to interview them, during the 2008 summer, by which time I was generally positioned as a non-threatening *niña* ('girl') in the neighbourhood (I was 26 when I first went into the field).

Six key *locutorio* users gradually introduced me to other users of the *locutorio*, more reluctant to participate in my research but all of whom, in the end, became part of the transnational *locutorio* network that I investigated. These six key informants were the people I most frequently met in the street: Jenny, an unemployed 35-year-old single mother whose family had come from southern Spain during the late 1950s; Shabbir, a 41-year-old construction worker who lived round the corner with his uncle, and who came from Kashmir; Sheema, a 31-year-old bar owner who arrived to town with his wife and two children from Pakistan; Abde, a 31-year-old undocumented and unemployed construction worker born in Morocco who came by boat; Merche, a babysitter and cleaner in her forties born in the Dominican Republic; and Nicolae, a documented 27-year-old factory worker (formerly self-employed) from Romania.

It was the *locutorio* worker Naeem, a 27-year-old Pakistani, though, who acted both as the most cooperative gatekeeper and as one of the most revealing informants whom I very soon befriended. From June 2008, he kindly gave me formal permission to carry out participant observation inside his workplace, and he helped me delimit the flexible boundaries of the transnational *locutorio* network. He provided me with key knowledge on this network's core and peripheral individuals, as well as on their life experiences, work projects and family roles, from his privileged insider's perspective as a person who, as a vendor, interacted with all of them almost on a daily basis.

Over a period of four months, I would sit, almost daily, near the *locutorio*'s desk and observe members of the network and simultaneously follow Naeem's movements during his 12–14-hour shifts, including at weekends. I left my bag where he left his bag; I ate with him and with other members of the network; I talked to the people they talked to (call card distributors, flatmates, and so on); I searched the same websites and listened to the same music; and I became familiar with the workings of the business. I was also taught about the technicalities of the place (for example, how much a call to Peru would cost, or what to do when a fax does not go through) and, in short, I got attuned to the minutiae of daily life at the *locutorio*. (Needless to say, I never got myself involved in money issues and I never managed the till!)

With a few exceptions, I was able to meet the informants on a regular basis for that and for the following year (systematically until August 2009), at the very least twice a week, for 2–12 hours. This allowed me to gradually learn about their relationships and their mobility plans, and to gain access to the *locutorio* network's many Goffmanian 'front-stage' and 'back-stage' discursive regions (Goffman, 1959: 22). I went around the neighbourhood with them, helped some Pakistani school children to finish some homework in Catalan, participated in an improvised cricket match in the street, frequently visited two other *locutorios* in a town nearby and shared a meal around Christmas time. I finished most days at the *locutorio* of El Paso.

Concerning the types of data that I was able to collect, unfortunately, I could not tape-record spontaneous interactions inside the *locutorio*, which would have been extremely valuable data with regard to the analysis of naturally occurring migrant–migrant interactions. Indeed, I was offered the possibility to do so without migrant interactants necessarily giving their informed consent, but I declined the offer. Even if I had posted a multilingual sign informing users that recording was in progress, many of the clients were undocumented, non-literate migrants, and some were involved in activities that were considered illegal. Others frequently engaged in extremely personal conversations, and the ethics committee at the UAB advised me not to involve myself in these. Besides, the *locutorio*'s owner, Naeem's superior, had sternly warned him and me, repeatedly, not to disrupt the daily workings of the *locutorio* (thinking that I was a member of the local police force, he once sent six Pakistani men to follow my moves past midnight, in the *locutorio* street). Therefore, I could not ask every client for informed consent, as this would have put Naeem's job at risk, which was a further limit of the data that could be gathered.

Nonetheless, I was able to triangulate several data sources. Firstly, I used the mainstay of ethnography, namely 'thick description' (Geertz, 1973: 310), which is an excellent tool for the investigation of sociolinguistic phenomena, particularly in the context of social inequality. Here, 'thick description' amounts to the gathering, via participant observation and an intense immersion into the *locutorio* world, of data that accurately describe systematised practices (communicative and otherwise) among transnational migrants.

The second set of data that I include in this book consists of 14 open-ended in-depth interviews involving 17 *locutorio* users and workers, all members of the transnational network that I followed, who gave me their oral informed consent to be recorded. These interviews were largely unstructured and occurred in a place of the informants' choice (normally on the benches or at the bars near the *locutorio*). They were conducted in the languages of the informants' own choice, although, due to my non-existent (or very limited) knowledge of their codes, these choices were mostly reduced to Catalan, Spanish and English, with Spanish clearly predominating as their choice. I learnt some basic Urdu (I was later told that in fact I was 'mixing' Urdu and Panjabi), but when I introduced myself to the Pakistani men of the network in this language, they were shocked and annoyed. I assume that they were uncomfortable with my learning of their 'we-code' (Gumperz, 1982: 66), because that was the language in which they conducted the activities that I was not supposed to hear about. In the end, then, we basically stuck to their preferred language option, Spanish.

On many occasions, the informants also decided upon the number of informants in the interviews. Thus, thirdly, I also have some data from interactional exchanges among migrants themselves. This proved useful, for instance, to demonstrate that, across migrant groups from different linguistic backgrounds, the language of interaction in El Paso is systematically non-standard Spanish.

In addition to these recordings, I also included some information gathered through non-recorded interviews which I conducted with two active members of this transnational group: Jenny, the woman born in town, and Lila, a young undocumented waitress from Uruguay. They both found the tape-recorder intrusive, unsafe and dangerous (and perhaps boring) and preferred me to write down their answers, which was also fruitful in terms of gaining new knowledge on the *locutorio* network that I studied.

A fourth source of data was a wide range of documentary materials: multinationals' leaflets and advertisements, international call cards, discount plans for international telephone services, room-for-rent advertisements, and the like. It was possible in many instances to gain permission to make a record of the songs that the informants downloaded at the *locutorio*, the ringtones of their mobiles, the CVs that they wrote, the electronic social security forms that they had to fill in and the text messages that they sent to each other. These are all included in this study as a crucial part of the research, too.

My field notes proved to be even more informative than might be expected. This is because informants would frequently contribute to their content by writing or drawing. In the end, I kept some paper for me and left some more on the desk, which in effect became the *locutorio*'s public notebook – at least for the literate. I ended up using scrap paper because my notebook reminded *locutorio* users of discouraging official administrative procedures, whereas scrap paper was more in tune with the *locutorio*'s life and was more inviting.

A final set of data did not come from the *locutorio* or even from the neighbourhood. This consists of an ethnographic look at the management of linguistic diversity and of migration within the Spanish telecommunications sector. I realised that I needed some data on the Spanish ICT market in order to better understand the linguistic barriers and market regimes migrants were responding to when they mobilised their transnational resistance resources in *locutorios*. The set of data that I gathered for these purposes included a record of the public language practices of the 30 ICT firms operating in Spain at the time of the fieldwork, between 2007 and 2009. These firms were: the three multinationals (Movistar, formerly Telefónica de España, Vodafone and Orange), 20 ICT ventures and Spanish-based start-up businesses, and seven recently launched migrant-oriented (i.e. 'ethnic') operators, which exclusively targeted migrant clients.

I analysed the languages that each of these 30 companies publicly employed in their call centres, official websites and advertising campaigns. I called the call centres of each company on at least three occasions, established a record of the languages spoken by the agents working there, and maintained contact with them via email as well. I also noted the languages used by each company on their websites, collected advertising leaflets and visited several of their phone shops in the Barcelona area in order to gather written information. This all helped me to construct a picture of the linguistic landscape of the Spanish telecommunications world that migrants encounter in Catalonia.

I also conducted six open-ended interviews (also with informed consent) with key ICT social agents (mostly people working for the three ICT multinationals), such as linguistic engineers, managers of mobile phone operations, translators and call centre assistants. I met and interviewed them about the Spanish telecommunications sector's management of migrant languages either at their workplaces or during my visits to the two most important events of the Spanish ICT world: the Mobile World Congress in Barcelona (12–14 February 2008) and the 7th Telecommunications Day in Catalonia, organised by COETTC (Col·legi d'Enginyers Tècnics i Pèrits de Telecomunicació de Catalunya, the Official Association of Telecommunications Engineers in Catalonia) (Barcelona, 30 September 2008).

Finally, I also draw on institutional documents, particularly from the Spanish government's Secretary of State for Telecommunications and the Information Society (SETSI), the Secretariat for Telecommunications and the Information Society in Catalonia (STSI) and the Observatory for the Information Society of Catalonia Foundation (FOBSIC), as well as some census information relevant for the analysis. I further draw on some archival research on the evolution of the Spanish telecommunications sector, in order to contextualise and historicise the emergence of *locutorios*.

A last word is needed concerning methods and data. In line with my understanding of the purpose of critical sociolinguistic research, and as part of my commitment to the informants who made this research possible, I place a strong emphasis on *ethnographic reflexivity* (Bourdieu & Wacquant, 1994; see also Heller, 2006 [1996], 2007a; Kroskrity, 2000; Sarangi & Candlin, 2001). Therefore, I include some autoethnographic data and 'personal narrative accounts' (Pennycook, 2012: 147) that let me show readers my approach to the data and to the people in El Paso. This means that I present my own positioning, perceptions and views in each of the interview excerpts of this book. Thus, we will find, for example, that I was taken up as *la catalana del barrio* ('the Catalan of the neighbourhood'), and that I participated in the mistrustful attitude towards the ICT multinationals that the members of the *locutorio* network had developed. I encourage readers to judge for themselves the types of questions that I asked and the many different roles that I played in the neighbourhood in order to emphasise that this is just my own particular situated interpretation of a slice of ethnographic experience for which I am solely responsible.

Overview of the Chapters

Having highlighted above the contextual information required to situate the *locutorio* phenomena within the context of the Catalan network society, in the second chapter I provide a historical perspective on the emergence of *locutorios* as unique migrant institutions of resistance and ICT-mediated subversive communication, and I analyse the rationale behind their unparalleled consolidation, in Europe, as truly alternative refuges and as powerful transnational meeting points. I first show that *locutorios* are testimony to a grassroots reaction against the top-down institutional barriers imposed on migrant populations by a hostile late-capitalist block consisting, in this case, of the Spanish nation-state and the Spanish telecommunications sector. I claim that, in complex ways, these two technologised apparatuses together exert a political hegemonic power upon migrant populations by gatekeeping their access to ICT via a series of legal measures, economic constraints and communication barriers. I conclude that these international regimes of late capitalism, which in Catalonia include the establishment of panopticon ICT dataveillance systems and old, ineffective 'multilingual' services mapped upon a marginalising Spanish-only monolingual floor, show that gaining access to citizenship today is about technology and, crucially, about language as well.

In Chapter 3, I detail the ways in which *locutorios* have become the real articulators of the migrant citizenry's successful self-insertion into their Catalan host societies, despite all sorts of adversities. I document the tactics by means of which current transnational family units and migrant individuals can find an alternative way to meet their basic needs in their own languages (including food, toilet services, discount plans for international calls, technoliteracy abilities, free legal advice, information on the social security system, and so on); particularly their strategies to collectivise and re-distribute ICT (SIM cards, internet access, handsets and the like). I pay special attention to the system brokers who actually make these tactics function – the overworked *locutorio* employees who, via their exhausting language work and their unrecognised intercultural translation and mediation tasks (ranging from the filling in of money transfer forms in Spanish to the dialling of given telephone numbers for the non-literate or non-numerate), uniquely endow their migrant clients with the social agency required for effectual ICT-mediated global trotting.

Chapter 4 starts with a critique of the ways in which the Spanish telecommunications sector, following the regimes and practices of the global market, imagines and presents migrants' ICT-mediated transnational communication practices. I provide evidence that the local Spanish market propels the emergence of unrealistic commercial identity labels (like the 'irresponsible mothers' or the 'mean friends' who never call) which publicise internationally sponsored homogenising and stereotyping images of migrant populations (and of migration phenomena). I then contrast this globally shared market-driven imaginary against the migrant families' *real* ICT-mediated communicative practices, as I observed them on the ground.

I also show that these transnational populations are, indeed, deeply immersed in the prevailing global capitalist consumer culture (particularly of goods like mobile phones). However, I argue, in turn, that they participate in the dominant clientelist regimes only with careful strategising with what multinationals and ICT companies offer them, and by developing tactics which subvert established local market rationalities. For example, I examine their use of 'missed' calls or of 'double SIM' mobiles, which allow for cheaper international calls without the need to sign a contract.

Finally in Chapter 4, I present some of the different ways in which local transnational family units organise their daily routines and fulfil their responsibilities via these communication 'tricks', at the *locutorio*. While I report some distressful experiences of particular individuals experiencing 'empty' conversations with their loved relatives and friends, I demonstrate that, in general, migrants actively manage to maintain intimate social ties by 'doing' proximity and by practising kinship regardless of physical distance, which provides further evidence that they have overcome their spatial dislocation and their technology 'have-not' position, in their own ways and through their own means, in the global information age.

Chapters 5 and 6 then examine the insides of the discursive spaces of *locutorio* shops, with a focus on situated local practices concerning language and identity. In Chapter 5, I present the sociolinguistic hierarchies and the intra- and inter-group competitions of linguistic capitals that regulate migrant-tailored *locutorio* worlds. Through a detailed examination of the linguistic features of the informants' xenoglossic and non-conventional room-for-rent advertisements, posters and texts messages, I argue that these institutions are transnational 'contact zones' (Pratt, 1991: 33) where many allochthonous codes meet with the two local languages, Catalan and Spanish, apparently establishing a seemingly multilingual territory. However, I show that these multilingualisms, in theory and in practice, get organised around, and mapped upon, a complex Spanish-regimented monolingual floor, which, in the end, not only silences or sanctions some of the migrants' codes (particularly non-national codes like Darija or Panjabi) but also delegitimises non-standard ways of 'doing' literacy (including technology-mediated grassroots literacies), to the detriment of everybody's allochthonous linguistic capitals, and to the detriment of the Catalan language. As a consequence, a fluid translinguistic (i.e. truly translocal) Spanish, which calques the dominant linguistic regimes that prevail in the metropolitan areas of Catalonia, but which is eminently rooted in hybridity and heteroglossy, is imposed upon the *locutorio*'s desk. This is a highly flexible common 'everyone's Spanish' that may act as a largely counter-hegemonic lingua franca, because, with its use (which has even penetrated the circuits of press capitalism!), migrant populations have appropriated at least some corners of the Catalan urban landscapes. However, I simultaneously analyse some of the informants' self-disciplining written practices in Standard Peninsular Spanish and, consequently, I maintain that the unification of discursive spaces in this unique polyphonic code paradoxically traps migrants into the reproduction of the local normativity regimes which work solely in and through the majority nation-state language of their host societies, thus confining many *locutorio* users to other-imposed and/or self-imposed linguistic marginalisation.

Chapter 6 examines the *locutorios*' disempowering side by presenting some of the unorthodox subject positionings and some of the defensive communicative tactics mobilised by these groups of excluded migrants. Through the lens of the migrants' self-attributed or other-ascribed rival identity affiliations and multifaceted derogatory social categorisation practices, I explore the discursively constructed pejorative labels they use within social networks inside the *locutorios*, such as *nenazas* ('sissies') and *boludos* ('jerks'). I suggest that these fluid social identities speak of intricate relationships (for example, among Pakistani and Moroccan men, or among young women from Latin America and Romania), which are based on simultaneous trust and distrust, friendship and enmity, and dominion and dependency. In the light of this, I argue that these migrant-governed alternative institutions, while clearly empowering for the *collectivity* of migrants, may also evolve into exclusionary spaces of transnational 'mismeeting' (Larsen & Hviid Jacobsen, 2009: 83) for individual migrants or given migrant social networks.

The migrant social players who best illustrate the subject positionings of the excluded are the *locutorio* workers who, challenged by the demanding clientele of the globalised new economy, and silently submitting to their superiors, are forced to navigate transnational living at the bottom of the social ladder. I conclude by analysing, on the one hand, the set of amazing self-defence language 'tricks' (like the emphatic repetition of imperative verbs or the circulation of white lies in Spanish) and, on the other, the presentation of various selves (as helpless pauperised workers, for instance) that, as a reaction to their workplace exploitation, they deploy in order to protect themselves from the harassment, entanglement and social stagnation which are produced and reproduced among migrants themselves.

Finally, by way of conclusion, I highlight what *locutorios* can contribute to in the opening up of unexplored arenas in the field of the sociolinguistics of globalisation, from an interpretivist, critical ethnographic perspective (Duchêne *et al.*, 2013). They here allowed the investigation of the under-researched mundane, translinguistic alphabets of the future that speak of the old, the new and the altogether different struggles, subversive techniques and resistance practices with which extremely different groups of migrants apprehend their transnational social worlds and successfully combat marginalisation by attaining 21st-century transnational survival.

2 The Rise of Anti-Migrant Governmentality: Prelude to the Emergence of *Locutorios*

> *To identify the owner of every telephone number is of crucial importance in our fight against terrorism and organised crime.*[18]
> Social Democrat Alfredo Pérez Rubalcaba, former
> Spanish Minister of the Interior (*El País*, 2009)

> *Obviously technology has nothing to do with languages; it is the same for everyone.*[19]
> Services and solutions consultant working for the multinational
> ICT company Movistar, in interview with the author at the
> Mobile World Congress, Barcelona, 12 February 2008

The aim of this chapter is to describe the social difference and marginalisation that migrants experience when trying to gain access to communication technology in the Catalan network society, particularly when trying to use mobile phones in combination with prepaid call cards, their transnational 'communication staple' (Panagakos & Horst, 2006: 112). I argue that in Catalonia there are two *technopolitical institutions of the global era* (that is, two governing bodies with privileged access to political, technological and economic power) with vested interests in the regimentation of migrant populations. Together they pose a series of largely novel exclusionary barriers to migrant participation. These two institutions are the Spanish nation-state and the Spanish telecommunications sector.

I begin by showing that such marginalising restrictions are part and parcel of the 'governmentality' that characterises the world's citizenship regimes in the 21st century – where 'governmentality' (following Foucault, 1979 and Pujolar, 2007b) is understood as the ways in which advanced neoliberal democracies control populations via a series of gatekeeping institutions and bureaucratic systems, according to the principles of logic, order and rationality, today with the crucial help of the private business sector, in coordination with financial lobbies (Castells, 2004; Fairclough, 2006; Harvey, 2003, 2005, 2010; Inda, 2006; Portes et al., 2001). I argue that these restrictions speak of the Spanish government's and the Spanish telecommunications sector's difficulties in navigating the complex dynamics of globalisation, which make them become crucial allies, but in a pugnacious race for accessing and maintaining *hegemonic power* (Gramsci, 1999 [1971]: 507) and cultural and political dominion both in Spain and globally. Thus, I explain how the Spanish government is strategically following the postsocial (cum increasingly market-driven) technologised citizenship regimes of leading European states in order to secure its political-economic place within supranational authorities, at a time when the telecommunications sector, one of the private institutions in Spain that has seemingly been able to resist the economic crisis (CMT, 2011), has also become involved in the policing of transnational citizens, through its access to one of the most lucrative segments of the market: migrant clients.

I then go on to examine the ways in which these two macro governmentality blocks exert their power locally. I first analyse the new dataveillance systems and the old covert linguistic regimes with which the nation-state regulates migrants' conditions of access to ICT, and then detail the sort of global capitalistic market rationalities and the ineffective form of 'commercial multilingualism' through which the Spanish telecommunications sector aggressively markets its services to them.

Against a background of migrant persecution, I conclude that both institutions make every effort to socialise and regiment migrants into a largely monolingual Spanish bureaucratic culture (Sabaté i Dalmau, 2013a). This attests to the fact that present-day exclusionary regimes are, above all, of a linguistic nature, for these anti-migrant policies, in practice, lead to the technological disconnection of those transnational migrants who are undocumented, pauperised or non-socialised into the prevailing dominant communicative regimes of host societies like Catalonia.

The Post-Social State: Exclusionary Dataveillance Systems and Covert Linguistic Regimes

While in the 20th century the Spanish government was characterised by a welfare orientation, it has in more recent times started to curtail its responsibilities to citizens and to become a market-driven, advanced liberal technopolitical institution. The central ministries, regional governments, local committees and agencies together act as one macro ruling body. Apart from using border patrols and repatriation systems, this 'technologised' governmentality block (Diminescu, 2008: 567) today plays out the real battle concerning undocumented citizenship in the realm of ICT (Ros et al., 2007). As the Spanish government controls telecommunications (article 149.1.21a of the Spanish Constitution; Parlament de Catalunya, 2007), it can regulate migrants' access to legal residence via the establishment of newer ICT dataveillance systems, following EU precepts. An example of this is the Law of Data Storage (Ley 25/2007 de Conservación de Datos). This was an adaptation of the ICT-mediated citizenship regimentation techniques established by leading North American and Asian states as a reaction to the terrorist attacks in New York and Washington DC in 2001 (O'Neill, 2003; Rheingold, 2002; Sahul Hameed, 2008). It was issued by the Spanish cabinet in 2006, following a European directive for ICT data storage approved by the European Parliament and the Council of Europe (Directive 2006/24/CE).

Ley 25/2007 obliged all operators offering landline and mobile telephony or internet services in Spain to register every client via an official identification document, starting as of 9 November 2007. Telephone companies were given six months to adapt their data storage systems to the new law, at their own cost. Unidentified users who had bought their phones before that date were obliged to show their Spanish identification card or their residence permit to their companies by 9 November 2009 if they wished to keep the same telephone number.

The rationale behind this bill, according to the then Social Democrat Spanish Minister of the Interior, Alfredo Pérez Rubalcaba, was to prevent further (supposedly migrant-organised) 'Islamist' terrorist attacks in Europe and, more particularly, in Spain.[20] In its preamble, Ley 25/2007 mobilised an internationalised 'national security' neoliberal rhetoric (Inda, 2006), stating that the way to secure Spanish cities was to control the conditions of access to ICT 'for the protection of people and goods and for the maintenance of public security', on the grounds that today they may easily be used for 'criminal purposes'[21] (*Official State Bulletin*; BOE, 2007: 42520). Emphasis was placed on the control of access to SIM cards, which at that time were being sold 'without control' (*El País*, 2009), particularly to migrants, 'whose initial preference is a prepaid phone which does not require the giving away of personal data or a bank account number'[22] (*CiberP@ís*, 2007: 1). Thus, the law (presented only in Spanish) targeted those who at that time had managed to skirt the gaze of the nation-state and the telecommunications sector by remaining unregistered: the migrants.

In theory, then, since November 2009 it has been illegal to own a mobile phone card without having shown proof of legal residence in Spain. The day after the deadline (9 November 2009), however, there were still around 800,000 unregistered mobile phones in Catalonia alone. The Spanish

Minister of the Interior, under pressure from the three big multinationals (Movistar, Vodafone and Orange), which had initially opposed the law and which were determined not to risk losing approximately €8–9 million if the disconnection of purchased SIM cards was to be carried out, then provided an extension of six months to the deadline (*Avui*, 2009). Many unidentified phone users (including most of the members of the *locutorio* network who participated in this study) had kept their mobile phones at the time of writing, but it is becoming more difficult to access a new line without registration, although less so in alternative institutions like *locutorios*.

I argue that Ley 25/2007 was one of the Spanish government's new advanced liberal *exclusionary migration policies*. The state's real aim concerning the targeting of undocumented migrants was strategically *implicit*: the word *usuarios* ('users') was used, so the government could not be accused of conducting an anti-migrant campaign. The way in which the law was presented, with the imperative '¡*Identifícate!*' ('Identify yourself!'; see Figure 2.1), together with the fact that it was issued by the Ministry of the Interior (notably, *not* by the Ministry of Industry, upon which telecommunications depend), provides further evidence that the Spanish state, following the EU, aimed to disconnect 'illegal' populations, in alliance with the telecommunications market, which agreed to comply with the law after some negotiation. This is thus the first legal barrier that transnational

Figure 2.1 The state's citizenship regimes and legal barriers. 'Compulsory registration plan for users of prepaid phone cards. Identify yourself!' (Ministerio del Interior, 2007)

migrants have to overcome when navigating the formal circuits of communication in Catalonia.

Framed in a context in which the Spanish state was following the European post-social paths to monitor citizens, Ley 25/2007 was presented as *una nueva manera de hacer Europa* – 'a new way to make Europe'. This was in fact the motto which, in 2008, the Spanish Secretary of State for Telecommunications and the Information Society (SETSI) adopted for a campaign known as *Plan Avanza* ('Plan Progress') in the digital management of migrant populations. This plan aimed to establish a model of economic growth based on increased Spanish competitiveness and productivity, via telecommunications (Area Moreira *et al.*, 2008). In agreement with the former monopoly provider and today's biggest multinational in Spain, Movistar, the SETSI campaign also claimed to facilitate access to ICT for populations deemed to be at 'risk of digital exclusion' (Plan Avanza, 2012), including those whom the government calls *nuevos ciudadanos* or *ciudadanos inmigrantes* ('new citizens' or 'immigrant citizens').

Figure 2.2 shows an image from the *Plan Avanza* campaign against migrants' 'technological exclusion'. It employs five short sentences in the first person singular (placed inside smiley codes) emulating the migrants' voice, monolingually represented in Spanish. They read as follows (from left to right): 'I request my EU citizenship certificate'; 'I check my foreign residence file'; 'I look for embassies and consulates'; 'I apply for a NIE [the identification number for foreign residents living in Spain]'; and 'I schedule an appointment at the Immigration Department'.

This campaign was launched as an index of technological leadership and bureaucratic modernity. It used rhetoric based on progressivist conceptions

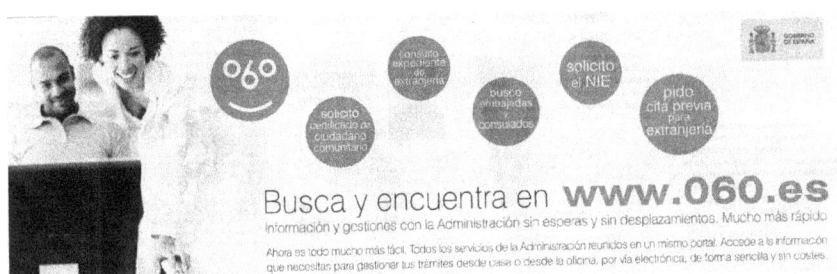

Figure 2.2 Image from the campaign against the 'digital exclusion' of migrants, *Plan Avanza* (*20 Minutos*, 2008: 23)

of citizenship and on a 'civic' (i.e. positive, welcoming) stance on migration. The Spanish governmental bodies claimed competence in the administrative management of migrants on the grounds that with this plan they were 'facilitating integration' and 'clarifying' the procedures for becoming a documented citizen, particularly for those who were unfamiliar with ICT and with the Spanish bureaucratic system. However, *Plan Avanza* did not actually meet transnational migrants' real needs. Instead, the ICT-mediated contact that the SETSI required concerning legalisation procedures in Spain entirely rested on the migrants' shoulders. The campaign urged them to first take full responsibility for their legal status and to engage in self-surveillance practices *before* being documented.

Plan Avanza achieved yet another governmentality goal: that of implementing an old but now largely covert linguistic regime or 'language testing practice' (Extra *et al.*, 2009) whereby the migrants who undergo legalisation via the new technologies are at present welcomed, socialised and disciplined in a largely habitualised monolingual written bureaucratic culture in Spanish only. The website through which to conduct the legalisation procedures online was entirely in Spanish in 2009. Today, some bits of information (basically, the headings, such as 'renewal of foreign national permit') are available in Catalan, Basque, Galician and English, but the actual information and all the necessary documents (e.g. temporary residence authorisations) are offered exclusively in Spanish.[23]

Thus, despite presenting itself as multilingual, the *Plan Avanza* in fact works in a monolingual manner, since the co-official languages in Spain are merely symbolic and peripheral, acting as a façade, while the migrants' codes are altogether absent. I suggest that this is the government's attempt to preserve a hegemonic unified floor in Spanish in a context of intensified linguistic diversity, again following other European states, which 'although nowadays *de facto* multicultural and multilingual, nonetheless still see themselves as essentially and indisputably monolingual' (Hogan-Brun *et al.*, 2009: 5), which bears witness to the fact that citizenship in Spain and in Catalonia is crucially regulated in and through language.

The Telecommunications Sector: Global Capitalistic Dynamics and Ineffective Commercial Multilingualism

As briefly outlined in the previous chapter, one of the macro techno-political institutions that has been gatekeeping the money flows and the accumulation of capital worldwide is the telecommunications sector (comprising multinational service providers, ICT operators, equipment manufacturers, terminal wholesalers, ICT enablers, retailers, and the like), which falls within the auspices of the private and the privatised economic spheres.[24] It is regulated by a global capitalist market, although in alliance (but also in competition) with nation-states, and has proved resistant to the economic crisis of recent times.[25]

At a local level, ICT gadgets have become a fundamental asset for the Spanish telecommunications sector, which has acquired political power through its involvement in the governmentality apparatus that was formerly hegemonically conferred to the Spanish government, particularly in managing the regimentation of transnational migrant clients, as briefly explained in the following section.

The rush to capitalise on the lucrative migrant niche

As happened in other 'post-industrial' countries, such as the United States (Fischer, 1992), migrant populations were for the most part absent from the history of the Spanish telecommunications sector, which began in 1924 with the birth of the Compañía Telefónica Nacional de España (CTNE). Today known as Movistar, the CTNE was nationalised in 1945, becoming the only telephone services provider and the 'natural monopoly' of the Spanish nation-state until 1995 (Puig et al., 2006), when, following the wave of liberalisation across Europe (detailed in Agar, 2003), new telephone companies (basically multinationals) were allowed to enter the sector for the first time (Fontela & Guzmán, 2000; Rozas Balbontín, 2003), propelling an impressive economic growth of 200% in the local mobile phone market (Boldó Gaspá et al., 1999: 28).

A second commercial boom occurred in 2006, when the European Commission conferred on the Comisión del Mercado de las Telecomunicaciones (CMT, the Spanish Telecommunications Market Commission) the power to force the three main operators (Movistar, Vodafone and Orange) to allow access to their networks to smaller emerging competitors, which would still dismiss migrants as prospective clients. These competitors were mostly mobile virtual network operators (MVNOs), based on the re-rental of the multinationals' lines. Their emergence brought the total number of operators in the Spanish state to 30 by 2009 (CMT, 2011).

Despite increased competition, though, the Spanish telecommunications sector still consisted of an *oligopoly*, which means that the three multinationals acted as an entrepreneurial block and enjoyed competitive advantages and better financial conditions (Guillén, 2005). Of the three, Movistar was the one which clearly dominated the market, to the extent that, today, it is still euphemistically called *el operador principal* ('the main operator') by the Ministry of Industry, creating the consumers' perception that the market is still essentially a monopoly.[26]

A third phase of market expansion can be identified within the Spanish telecommunications sector, when it nearly reached saturation levels (ONTSI, 2011). At this time, the number of mobile phone numbers registered per 100 inhabitants reached 112 (ITU, 2012b: 195),[27] propelling yet a fourth phase of increased competition among companies. ICT ventures operating in Spanish territory then started to seek 'distinction' (or added-value) and to make their business models more flexible in order to find new market niches (i.e. new groups of consumers). They often did so by lowering their flat-rate prices (CMT, 2011) and, of course, through language, via the establishment of apparently more 'multilingual' customer services, which had become a sign of innovation and competitiveness (Duchêne, 2011; Kelly-Holmes, 2005; Kelly-Holmes & Mautner, 2010).

It was during this fourth phase that transnational migrant populations decisively came into play and were finally conceived of by ICT ventures as an important economic niche (Ros & Boso, 2010). Telephone companies quickly realised that migrant populations were then already spending 40% more on telecommunications than non-foreigners (Robledo, 2008), and that they were generating an average income that, according to multinationals like Orange – which now had an 'immigrant customer services' department – was 'much higher than that which a Spanish client can offer'[28] (*CiberP@ís*, 2007: 1).

In their race to capitalise on migrants' pockets, the three multinationals started contracting with specialised market research companies in order to target this particular clientele in their marketing. In 2007, for instance, Nielsen started investigating migrants' ICT uses for Orange and for Vodafone (Nielsen, 2007). Likewise, the Asociación para el Conocimiento de la Población Inmigrante (ACPI, Association for Information about Immigrant Populations), whose corporate aim is to provide 'objective, validated market information about a niche, that of immigrants, which is increasingly gaining strategic importance'[29] (ACPI, 2007: 2), started working for Movistar (ACPI, 2008).

Out of the 30 ICT businesses that I could identify in 2009, seven were new MVNOs which presented themselves as 'ethnic' operators, that is, as exclusively migrant-oriented mobile phone ventures (see Table 2.1). These were all launched only after the 2006 market expansion, on the initiative of foreign investors, and with the help of the three multinationals, which now provided them with network infrastructure in a disguised manner, under more migrant-appealing commercial names written in the majority nation-state languages of their targeted clients, like the Orange-dependent Hong

Table 2.1 Migrant-oriented mobile phone operators in the Spanish state (2009)

'Ethnic' MVNOs (2007–09)	Year licence first granted	Network provider and investors	Commercial presentation
Happy Móvil	2006	Orange, The Phone House	'The mobile to call your country' ('El móvil para llamar a tu país')
Lebara Móviles	2007	Vodafone	'The best way to call your country from your mobile in Spain' (presented in English)
Talkout Móvil	2008	Orange, Euphony España, E-Plus Móviles	'A mobile telephony service specially designed for foreigners' ('Un servicio de telefonía móvil diseñado especialmente para extranjeros')
MundiMóvil	2008	Orange, no Telecom, E-Plus Móviles	'The first mobile phone company of the world which aims at overcoming communication barriers' ('la primera compañía de telefonía móvil del mundo que pretende romper las fronteras en las comunicaciones')
Hong Da Mobile ('Strong, powerful Mobile')	2008	Orange, Netia, E-Plus Móviles	'Your mobile to call China' ('Tu móvil para llamar a China')
Digi Mobil ('Digital Mobile')	2009	Movistar, Digi Spain Telecom	'Specially designed for Romanians living in Spain' ('[...] pensado especialmente para los rumanos que están en España')
LlamaYa Móvil ('Call Now Mobile')	2009	Orange, More Minutes Communication S.L.	'The lowest rates for all your calls'

Source: Official websites, leaflets, and personal electronic and telephonic communication with enterprises. Data assembled by author. Companies are presented by year of license granting, in increasing order

Da Mobile ('strong, powerful') MVNO sponsored in Mandarin Chinese, or the Romanian Digi ('digital') Mobil, set up in partnership with Movistar.

The emergence of both market research companies and ICT service providers specifically targeting migrants as a way to avoid market stagnation speaks of a 21st-century *clientelisation* of transnational migrant populations on the part of the Spanish ICT sector. This simultaneously propelled, in Catalonia as elsewhere, the *commodification* within the private sphere of a homogenised social category, that of *el segmento inmigrante* ('the immigrant segment') (*CiberP@ís*, 2007: 6). This commodification is modulated through a private-sector neoliberal discourse of 'integration', as exemplified, at a local level, in one of the most important annual business events in Spain: the Madrid Products and Services for the Immigrant Fair, interestingly called Integra Madrid ('Integrate Madrid' or 'Madrid integrates' – it can be read both ways). Since 2007, this has been one of the most important commercial meeting points for more than 150 different companies – particularly multinationals (Movistar, Vodafone), MVNOs (Digi Mobil, Happy Móvil, Lebara Móviles, LlamaYa Móvil) and financial service providers (La Caixa, Banco Santander, Banco Popular, Western Union) – to sell their products to migrant-regulated associations and NGOs, with the governmental support of the Spanish Secretary of State for Immigration and Emigration (IFEMA, 2008; ParaInmigrantes.info, 2009, 2011).

Presented as an excellent platform 'for the promotion of the *integration* [...] and *communication between the business sector and this specific segment of the population* [the migrants]'[30] (ParaInmigrantes.info, 2007, my emphasis), the Fair illustrates the regimes of thought with which the Spanish telecommunications market currently approaches migrant phenomena, which are based on a neoliberal celebratory 'unity in diversity' discourse, where diversity is constructed as an emblem of social unity and cohesion (Heller, 2006 [1999]; Jaffe, 2007; Muehlmann & Duchêne, 2007). What is perhaps new in these discourses, though, is that such 'cohesion' necessarily incorporates a phase of consumerism into the social 'integration' process, as if purchasing mobile phone telephony was a *rite of entrance* and admission into the host societies. Thus, under late capitalism, migrants are basically seen as properly socialised only after they have become '*Homo consumentus*' (Lindgren et al., 2002: 28) or fully fledged autonomous transnational capitalist selves who have learnt to make choices in the global marketplace. And what is even newer is the extent to which this sector has attained the power to regiment migrant populations, as detailed below.

The regimentation of migrants by the private sector

The first obstacle that the Spanish telecommunications sector poses to transnational migrants is of a legal nature and has to do with the difficult route to obtaining a SIM card in Catalonia. When ICT companies realised that the Law of Data Storage would not lead to a substantial loss of clients, they soon enough aligned with the government on that matter. Since 2009, ICT companies in Spain have had to provide a record of all their clients (a *libro-registro* or 'book-register') to the authorities; in turn, therefore, they require evidence of legal residence before they issue telephone contracts. This makes it more difficult for migrants, and especially undocumented migrants, to gain access to telecommunications services. When it comes to dealing with big multinationals, it makes it very difficult, too, for those *documented* non-Spanish nationals who do have temporary or permanent legal residence.

Perhaps fearing unpaid debt – GSM Spain (2009) reported that the number of unpaid telephone invoices increased by 57% during 2008 – some multinationals decided to take a step further and require new migrant clients not only to be documented but also to show a work permit or, in some cases, a Documento Nacional de Identidad (DNI, the ID card for Spanish nationals), for the provision of a telephone contract. As Nicolae, a documented 27-year-old (formerly self-employed) factory worker from Romania, put it in one of our interviews: 'Not even as self-employed, nothing! They [multinationals] don't even even want to see the papers [legal foreign residence] ... they want to see the DNI.'[31] Indeed, the need to be a Spanish national was asserted, unofficially and off the record, by individual agents working for that specific company, Movistar, in one of their call shops in the metropolitan area of Barcelona. Thus, the private business sector, in synergy with the nation-state, seems to participate in the machinery of an exclusionary citizenship regime through legal and commercial barriers which control who is granted full citizenship rights and who is not, via technology (see also Sabaté i Dalmau, 2013a: 252–261).

In addition to this legal constraint on migrant participation in the ICT market, a second constraint is of an economic nature. This is especially evident in the case of the multinational Movistar, which is often playfully referred to as *Timofónica* ('Swindle-phone-ica') or *Robofónica* ('Rip-off-phone-ica') in the neighbourhood of El Paso. The main problem that migrant clients face is that the bureaucratic processes for signing a contract (and, in fact, the details of the written documents themselves) are not always fully understood, accessible or transparent, which often results in misunderstandings about whether the VAT is included in quoted prices, about the payment terms or about the binding conditions of the contract. Thus, many migrants are very suspicious of the large service providers because of things like the amount to be paid not matching what they believe they had agreed to pay, unexpected charges for late invoice payments and the imposition of 12-month or 18-month binding contracts, as well as the difficulties of cancelling signed agreements. In short, the profit-securing economic regimes established by ICT enterprises in Spain are not geared towards the migrants' needs. Rather, they seem to further complicate their access to telephony services, which holds many migrant users back from signing contracts with big ventures, leading them to choose prepaid SIM cards and to prefer alternative providers of ICT services – like *locutorios*.

How these market rationalities can be perceived as a serious alienating economic constraint by transnational migrants is suggested in Excerpt 2.1, where Abdelouahed ('Abde'), a 31-year-old unemployed man born in Morocco, emotionally explains how he ended up receiving unexpectedly high bills after a late invoice payment. He consequently decided to forget about contracts and to go back to the use of the prepaid cards which were readily available in call shops like *locutorios*.

Excerpt 2.1. Capitalist market rationalities and economic constraints: Migrants' mistrust of multinationals

@Location: 1 September 2008. Bar near a *locutorio* in El Paso. Vallès Occidental.
@Bck: Abde (ABD) tells the researcher (RES) about an unexpectedly high bill he received after a late payment to a multinational.

```
  01  *RES:  y <tu> [/] <tu móvil es de tarjeta o de contrato> [¿].
      %tra:  and <your> [/] <your mobile is a top-up or contract> [¿].
  02  *ABD:  no de recarga # yo no me gustan los contratos.
      %tra:  no a top-up # me I don't like contracts.
  03  *RES:  <lo probaste> [¿] <con qué compañía> [¿].
      %tra:  <you tried it> [¿] <with what company> [¿].
→ 04  *ABD:  lo probado con Vodafone me sale bien y con Movistar me
             salido mal.
      %tra:  tried it with Vodafone it went well and with Movistar went
             wrong.
  05  *RES:  es más caro¿
      %tra:  is it more expensive¿
→ 06  *ABD:  es más caro pero más chungo también.
      %tra:  it's more expensive and dodgier as well.
```

```
     07  *RES:   sí¿
         %tra:   really¿
→    08  *ABD:   Movistar te # por ejemplo digamos tú tienes no tienes dinero
                 y luego la factura tarda dos días una semana y se ha doblado
                 la factura # Movistar a mi me dobló.
         %tra:   Movistar they # for example say you have you don't have
                 any money and then the bill takes two days a week and the
                 bill has doubled # to me Movistar doubled it.
     09  *RES:   sí¿
         %tra:   really¿
→    10  *ADB:   a mi me ha doblado una vez!
         %tra:   it doubled it once to me!
```

Abde searched the market and tried the mobile phone services of two multinationals: with one, Vodafone, it went well but with the other, Movistar, it did not (line 4), for he found himself having to pay extra charges when he did not have enough to pay his monthly invoices, as he very emphatically states through repetitions (in lines 8 and 10). This made him go back to the use of prepaid cards, by means of which he can control his expenditure on a daily basis (as opposed to simply being invoiced at the end of the month). This experience with the Spanish market regimes convinced him that those companies did not fulfil his needs and were in fact beyond his particular economic means for establishing ICT-mediated communications. Thus, he, like most migrant users of *locutorios*, developed a mistrust of the biggest multinational, which he now finds not only expensive and difficult to deal with but also untrustworthy and dangerous for clients like him (line 6).

A third structural barrier that migrant populations encounter consists of the Spanish telecommunications sector's established linguistic regimes. Basically, ICT companies offer a particular kind of multilingual communication technology which is mostly ineffective and which poses a serious challenge for many transnational migrants, because it neither fully understands their linguistic practices nor fulfils many of their communicative needs, for three main reasons. Firstly, the sector is clearly rooted in, and mapped onto, a Western *written culture*. Secondly, this segment of the private business sector is also mapped onto a *Spanish-only* linguistic regime, which in the end provides a largely monoglossic unified floor, despite some publicised welcoming rhetoric around 'linguistic diversity', circulated during leafleting campaigns. Thirdly, on the few occasions when the sector does make use of languages other than Spanish, the offer tends to consist of only a narrow set of dominant languages (like English) which are basically oriented towards an already connected 'global elite' (Bauman, 2005: 15). In this linguistic landscape non-elite allochthonous codes are excluded or at best rarely found in this sector's fake 'multilingual' repertoire. Its particular type of multilingualism is aptly referred to as *commercial multilingualism*, that is, 'a reflection of and a desire to respond to "real" multilingualism in society, particularly where the speakers of languages are seen to be economically attractive targets', which, in fact, 'may have very little to do with lived multilingualism' (Kelly-Holmes, 2010: 23).

In what follows, I analyse this commercial management of linguistic diversity, and I elaborate on the linguistic burdens that it poses to migrants, particularly to those who are non-schooled, non-literate or non-numerate in the prevailing literacy and numeracy practices of the northern hemisphere. By drawing on some previous work (Sabaté i Dalmau, 2012a, 2013a), I first look at how the dominant Western written cultural practices emerge as a literacy burden for migrant populations, even for literate ICT clients in particular contexts involving the written use of both autochthonous and allochthonous languages. I then analyse the linguistic ideologies and the discursive tropes on linguistic diversity mobilised by the Spanish telecommunications companies, and try to understand the rationale behind them. Finally, I focus on the companies' *real*, habitualised linguistic practices in service delivery, by investigating the languages in which the 30 ICT ventures publicly commercialised their mass consumption services and products in Catalonia, at the time of the fieldwork. These languages basically include Spanish, English (and other dominant nation-state European languages like French and German), Catalan and, finally, non-elite allochthonous codes like Modern Standard Arabic or Romanian (presented below in this order).

The literacy regimes of the dominant Western written culture

The sort of homogeneous written culture around which the global ICT system is organised makes it difficult not only for non-literate but also for literate migrant clients to have ready access to some basic technology services. This is explained by Abde in Excerpt 2.2, where he complains about the fact that he cannot make use of text messaging services, the cheapest option for exchanging short bits of written information via a mobile phone, when contacting his partner Aisha in Morocco, due to their incompatible linguistic capitals in the written mode.

Excerpt 2.2. Literacy regimes: A communication system based on the written mode
@Location: 28 August 2008. Bar near a *locutorio* in El Paso. Vallès Occidental.
@Bck: Abde (ABD) tells the researcher (RES) about the impossibility of using SMS services with his partner Aisha in Morocco due to the incompatibility of their linguistic capitals when establishing written ICT-mediated contact.

    01 *RES: mensajes mandas¿
           %tra: do you send messages¿
→ 02 *ABD: no.
        *RES: no¿
→ 03 *ABD: para ella no le mando mensajes porque <yo> [//] ella no va a leer por castellano # ella puede leer árabe pero yo no le puedo escribir árabe -, yo tampoco fui a la escuela a mi país.
           %tra: to her I don't send messages because <I> [//] she won't read Castilian # she can read Arabic but I can't write Arabic to her -, I didn't go to school in my country either.
           %com: Spanish is usually called Castilian in Catalonia, in reference to the region of the Spanish state where it comes from.
    04 *RES: ah vale.
           %tra: ah ok.
→ 05 *ABD: yo si escribo escribo castellano.
           %tra: me if I write I write in Spanish.

Abde does not send messages to his partner in Morocco (line 2) because he is able to manage the written modality only in Spanish (line 5), a language which Aisha cannot read or write (line 3). By contrast, she can read and send SMSs in Darija and in Modern Standard Arabic, but since Abde did not go to school – which is the case for about 40% of migrant men of Moroccan origin in Catalonia (Hernández-Carr, 2011: 99) – he cannot access these languages in written form (line 3). This is why, he says, they have not tried using written French either. Thus, with his partner there is actually no other way but to call her. This is a type of difficulty which illustrates that even literate technology users may encounter problems when engaging with a global ICT communication system deeply nested in, and geared almost exclusively towards, written modes of communication.

A 'multilinguistic model' rhetoric

As a strategy to welcome transnational migrant clients, the Spanish telecommunications sector today makes use of a positivising global rhetoric linking linguistic diversity to social inclusiveness, with discourses which allow most companies to attribute to themselves (and publicise) what Tan and Rubdy (2008: 2) call entrepreneurial 'multilinguistic models'. This rhetoric belongs to the dominant neoliberal discursive repertoires which associate an undefined 'multilingualism' with a high degree of innovation, efficiency, competitiveness and modernity, as well as of civility, acknowledgement and respect for ethnolinguistic and cultural diversity (Duchêne, 2009, 2011; Heller, 2006 [1999]; Jaffe, 2007).

The self-attributed use of 'multilinguistic models' has become the prevailing marketing leitmotif exploited by the vast majority of ICT companies in the Spanish telecommunications sector (and elsewhere, too), particularly by those which are migrant-oriented. This is highlighted, for instance, by the Spanish corporate media, which now present the main characteristic of this segment of the market as having become 'highly multilingual', with headlines like 'Everyone in their own language. This is the key to reaching the foreign community in Spain' (*Metro*, 2008a: 15).[32] The comments made by a services and solutions consultant from Movistar in our interview at the Mobile World Congress in Barcelona epitomise this welcoming discursive trope and illustrate my claims in more detail, in Excerpt 2.3.

Excerpt 2.3. Commercial multilingualism: A marketing leitmotif

@Location: 12 February 2008. Mobile World Congress. Movistar stand. Barcelona.
@Bck: A Movistar services and solutions consultant (CON) tells the researcher (RES) about the company's 'multilinguistic model' and its corporate commitment to offering multilingual services (basically, in Spanish and English as global linguae francae).

```
01 *CON: somos multilingües # la convergencia también obliga,,
         <no> [?] obliga a abrir una plataforma multilingüe -, los
         servicios van a tener que tener la posibilidad de ser multi-
         lingües porque sino a lo mejor no voy a poder comunicarme
         con alguien que está al otro lado.
    %tra: we are multilingual # convergence also obliges,, <right> [?]
         it obliges [us] to open a multilingual platform -, the services
         will have to have the possibility of being multilingual
```

		because otherwise perhaps I won't be able to communicate with someone who is on the other side.
02	*RES:	ah.
→ 03	*CON:	con el inglés va a ser más posible,, <no> [¿] porque tanto el inglés como el español son idiomas potentes son fuertes extendidos en muchos países # además el idioma oficial de Telefónica es el inglés.
	%tra:	with English this will be more plausible,, <right> [¿] because both English and Spanish are powerful languages they are strong spread in many countries # besides the official language of Telefónica is English.

Excerpt 2.3 shows that multinationals in the Spanish ICT world have embraced the rhetoric of 'multilingualism as added-value', but that they have done so as a kind of corporate acknowledgement of a general, abstract or loosely defined, 'empty' linguistic diversity (Krzyzanowski & Wodak, 2011: 126). This fashionable floating label of late capitalism indexes Europeanism and inclusive post-nationality, and allows companies to present themselves as belonging to a global information era where linguistic barriers and communication problems are imagined to be overcome.

Excerpt 2.3 also exemplifies that the 'multilingualism' that is offered is often an *elite multilingualism* suiting the needs of a technoliterate elite (usually, the tourist class or the transnational upper-middle entrepreneurial class). The services are offered mostly in a set of 'dominant languages' which mobilise huge investments and flows of capital worldwide and which play an internationally important role in globalisation processes (Heller & Martin-Jones, 2001). Thus, in the Spanish telecommunications world, many ventures present themselves as 'multilingual' simply on the basis that, as the Movistar consultant's observations show (in line 3), they promote the use of dominant languages like English and Spanish as the two most frequently used 'strong' 'powerful' linguae francae, that is, as a precondition not only for international business transactions but also for successful intercultural communication within the Spanish state.

The non-management of the migrants' multilingualisms

Against this commercial rhetoric and this self-attributed multilingual competence, a systematised in-depth analysis of the public language practices of the 30 mobile phone ventures operating in Barcelona reveals that, for the telecommunications sector, 'being multilingual' means offering tokenistic, inconsistent, non-useful information and few services only in

European majority languages, in a secured Spanish-unified marketplace. This is summarised in Table 2.2, which presents the linguistic landscape of the Spanish telecommunications sector concerning its customer service call centres, where clients get to talk to company agents (the data are taken from the work that I presented in Sabaté i Dalmau, 2012a: 144).

Spanish

The linguistic landscape indicated in Table 2.2 shows that companies basically offered the vast majority of services in and through *Spanish as the default language* or, in some cases, as the only language, in their call centres. Besides, out of the 30 ICT ventures that I could record during the time of the fieldwork, the vast majority (20 start-ups and smaller MVNOs) used Spanish as their *only* language on their websites and in their advertising campaigns, except for the Basque company Euskaltel Móvil, the Galician R Móvil and the Catalan RACC Mòbil, which operated respectively in Basque, Galician and Catalan as well as in Spanish.

There are several reasons for the institutionalisation of this habitualised Spanish-unified market, which has often been explicitly funded and supported by successive Spanish governments (Del Valle, 2006; Mar-Molinero, 2006). The first is that, in the globalised new economy, pragmatism dictates that there has to be an 'economisation' of linguistic diversity (Duchêne, 2011: 102). From my point of view, this means that the languages offered to clients are those deemed a profitable resource. This is in line with a 'discourse of profit' (Heller & Duchêne, 2012) or of 'linguistic instrumentalism' (Wee, 2008) which primarily treats language in economic terms and presents it as a marketable consumer item devoid of social, cultural or political meanings, to be sold or purchased in an utilitarian manner when pursuing economic development and expansion (Alarcón, 2011; Duchêne, 2009, 2011; Heller, 2003, 2010b; Rovira Martínez *et al.*, 2005).

Following this 'economisation' of linguistic resources, Spanish in the Spanish telecommunications world is constructed by the multinationals' language economists[33] as the 'natural', default language that meets the needs of various local populations living in a monolingually depicted, uniform Spanish nation-state. Spanish is presented as a de-ethnicised (and profitable) lingua franca which may be highly instrumental for the entire 'Spanish-speaking condominium' (García Delgado *et al.*, 2007: 21) (i.e. for targeting the Spanish-speaking consumers in Latin America), at a time when companies like Movistar have expanded to 13 new countries where Spanish is a (co)-official language. Consequently, the Spanish language in the ICT world is now frequently talked of as a 'multiethnic language', a *lengua sin patria* (a 'language without a fatherland') or a 'post-national

Table 2.2 The linguistic landscape of the telecommunications sector in Barcelona, 2007–09: Customer service call centres

Company	Languages offered
Movistar	Spanish, English, Catalan, German, French, Polish, Romanian, Arabic
Vodafone	Spanish, English, Catalan, German
Orange	Spanish, English, Catalan, Basque, Romanian, Arabic, Galician
Bankinter Móvil	Spanish, English, German, Catalan
Blau	Spanish
BT Móvil	Spanish
Carrefour Móvil	Spanish, English
Digi Mobil	Romanian, Spanish
Día Móvil	Spanish, English
Eroski Móvil	Spanish, Catalan, Galician, Basque
Euskaltel Móvil	Spanish, Basque
FonYou	Spanish
Happy Móvil	Spanish, Russian, Romanian, French (English, Arabic)
Hits Mobile	Spanish, English, Dutch
Hong Da Mobile	Not operational
Jazztel	Spanish (English)
Lebara	Spanish, English, French, Russian, Portuguese, Italian, Arabic, Dutch, Polish (Catalan, Galician, Lithuanian, Urdu)
LlamaYa	Spanish, English
MundiMóvil	Spanish, English
Más Móvil	Spanish, English, German
Pepephone	Spanish, English, Catalan
RACC Mòbil	Spanish, Catalan
R Móvil	Spanish, Galician
Simyo	Spanish
Sweno	Not operational
Talkout	Spanish, English
Telecable	Spanish
Vueling Móvil	Spanish, English, French, Dutch, Italian, Catalan
XL Móvil	Spanish
Yoigo	Spanish (Catalan soon)

Source: Data assembled by author, drawing on the corporate information provided by each company via electronic or telephone communication. The order in which languages are presented is the order given by call centre agents. Languages in parentheses were offered non-officially, on the agents' own initiative

language', an 'anonymisation' and 'deterritorialisation' positioning which is convenient for its economic and political exploitation (Del Valle, 2006, 2007; Mar-Molinero, 2006; Pujolar, 2007c; Woolard, 2006).

In these profitability discourses, the policiticised defence of a unified linguistic market is presented as an integral part of the necessary defence of a Spanish telecommunications sector in times of crisis, both at a local and at an international level. The language economists from Telefónica Foundation, for example, emphatically call for an economisation of linguistic resources (i.e. the sole use of Spanish) as the way to secure the position of the Spanish business sector in the face of increased competition from multinational businesses that use linguae francae such as English (García Delgado *et al.*, 2007; Martín Municio *et al.*, 2003).

This does not mean that the older ethnicist or culturalist discourses that link language with national identity and national pride have altogether been abandoned. Rather, it means that nativist conceptions of language (traditionally mobilised by governmental apparatuses constructing a national unity) have been reinscribed in these economicist discourses in very complex ways (Heller & Duchêne, 2012: 1–6). The tremendous political and cultural machinery that has been mobilised to boost the spread of Spanish as *the* global language of the ICT world, for instance, is an example of how companies with a Spanish nationalist agenda have carefully modulated their rhetoric at a time when it is more difficult for them to overtly project a staunch linguistic militancy (Pujolar, 2007a). Microsoft and the Real Academia Española (RAE, Royal Spanish Academy, a purist Spanish language regulatory organ), for example, explained their 1999 agreement to incorporate 'the best possible Spanish' in Microsoft's computer programs as a way to help the language navigate the road of global digitalisation in the 21st century. Rodríguez Lafuente, the then-director of the Instituto Cervantes, whose aim was 'to win the language battle', signed a similar agreement with Telefónica, which he justified in terms of surveying the presence and the 'economic weight' of Peninsular Spanish on the web (Del Valle & Gabriel-Stheeman, 2002: 196, 206–207).

English

The ICT businesses which initially operated only in Spanish in Catalonia soon found they had to take some account of linguistic diversity when managing at least their face-to-face customer services, in order to avoid losing clients (Alarcón, 2007, 2011; Barcelona Activa, 2011; Marí & Strubell, 2011; Solé & Alarcón, 2001). As a consequence, today, within the largely monoglossic Spanish market that I have described above, several companies do offer some of their products and services in other languages, albeit these are largely (but by no means systematically) the globally dominant ones, notably *English as the global language of technology* and as the language for *international business management*, but also, to a lesser extent, French, German and Dutch and, just in one case, Italian. Thus, in 2009, half of the 30 operators offered customer services in English (see Table 2.2).

I argue that English is the second code in the Spanish ICT market because, in the international arena, it is the language employed by global think-tanks and powerful governmental agencies (Fairclough, 2006: 6), such as the International Telecommunication Union (ITU), the World Trade Organization (WTO) and the World Bank (WBG):

> The high status accorded to English, especially Standard English, has been reinforced by the significance it has assumed in the global cultural economy. English now represents the preferred language medium in which most transactions take place within transnational business organizations, as well as in political or economic encounters such as those that characterise the World Trade Organization, World Bank and Monetary Fund ensembles. (Tan & Rubdy, 2008: 6–7)

While clearly rooted in a Spanish-using entrepreneurial machinery, the Spanish telecommunications sector has started to sponsor the use of English as a way to help ICT companies navigate the convoluted globalised new economy, with campaigns in this language promoting Spain's technological leadership. These campaigns include mottos playing with the English language, like 'España, technology for life' (used at the Mobile World Congress, 12–14 February 2008, Barcelona). The front-stage use of English by this sector is a strategy to maintain the leadership and the technopolitical position it has enjoyed, for instance as host to the annual Mobile World Congress, the biggest event of its kind, in Barcelona, since 2008. English, as the taken-for-granted language of business interactions, dominates the circuits of commercialisation regulated by an elite expert polity, as the chief staff manager of the Mobile World Congress held in Barcelona in 2008 explains in Excerpt 2.4. Notice that the supremacy of English remains unquestioned and is repeatedly highlighted (in lines 3, 5 and 7).

Excerpt 2.4. Going global: English as a lingua franca for the technoliterate elite

@Location: 12 February 2008. Mobile World Congress. Chief staff manager's room. Barcelona.
@Bck: The researcher (RES) asks the chief staff manager (STA) about the languages being employed at the Mobile World Congress, and she emphasises the overarching presence of English in this event.

01 *STA: preguntan por Telefónica [Movistar] # preguntan mucho
           también por Google # um por Vodafone -, bueno las grandes.
   %tra: they ask about Telefónica [Movistar] # many ask about
           Google a lot too # um about Vodafone -, well the big ones.
02 *RES: las grandes vale eh y: normalmente <en qué lenguas lo
           preguntan> [¿].
   %tra: the big ones ok eh a:nd normally <in what languages do
           they ask about them> [¿].
→ 03 *STA: inglés.
   %tra: English.
04 *RES: inglés¿
   %tra: English¿
→ 05 *STA: inglés el noventa por ciento de las preguntas son en inglés.
   %tra: English ninety per cent of the questions are in English.
06 *RES: sí¿
   %tra: really¿
→ 07 *STA: sí en inglés y si no castellano pero es raro.
   %tra: yes in English and if not then in Castilian but this is unusual.

Also, global English is the default model language in which to try new technological developments. This is true not only for multinational companies outside Spain – for example, Samsung, LG and Philips are moving towards an English-only policy (Tan & Rubdy, 2008: 1) – but also for powerful, well established Spanish ventures. For example, the Orange voice assistant launched in Spain was first commercialised in English, later in French and finally in Spanish (Connectta't, 2008). The same occurs with the vast majority of translation and voice recognition programs, which are developed in English and later transcoded into other languages.[34]

This taken-for-granted use and prestige of global English ('the official language of Telefónica is English', stated the Movistar's representative in Excerpt 2.3 above) explains why it is basically the multinationals – Movistar, Vodafone and Orange (and not the smaller, local start-ups) – which offer English in their call centres, though still not consistently. Moreover, customer services in English are preferentially offered to privileged clients belonging to the technological elite (the company Jazztel, for instance, offered English only to Jazztel investors, not to all clients or to non-Spanish-speaking migrants, at the time of the fieldwork). This indicates that, in fact, contrary to what many telecommunications companies claim, English has by no means become the *real* lingua franca of the entire Spanish ICT realm, but just an unsystematic second language choice. The fact that only four out of seven migrant-oriented operators offered English systematically in their call centres in 2009 (Lebara, MundiMóvil, Talkout and LlamaYa, the last only on weekdays) provides further evidence for these claims – a fifth 'ethnic' MVNO offered it after 4 pm unofficially, on one of their individual agents' private initiative. In short, many of the ICT companies targeting migrant clients claim to have a 'multilingual competence' geared by and through the presentation of English as an adequate second choice but, *de facto*, end up managing linguistic diversity, once again, in Spanish, as the following ethnographic snapshot, taken from my field notes (on 21 September 2008), illustrates (Snapshot 2.1).

Snapshot 2.1. Problematising the role of English as a lingua franca for migrants in the Spanish ICT world

Rachid, the 20-year-old Pakistani who has been Naeem's *locutorio* trainee for three days now, wants me to call Orange and ask the company to send him their customer service messages indicating the credit remaining on his mobile phone in English, instead of Spanish. During his two-year stay in Greece, he says, he also used prepaid services with this multinational, which apparently offered him the possibility to receive SMSs in either English or Greek (he chose English, a co-official language of Pakistan along with Urdu). This is why he is now assuming that in the Spanish state he will also have the choice between Spanish as the matter-of-fact nation-state language (for which he says, in English, that he is not ready), and English as the global language, which would make SMSs intelligible for him. I suggest that he tries it himself, but end up making the call from Rachid's mobile. I just ask directly whether there is an English language choice for company-to-client SMS communication. Baffled, the call centre worker replies that he regrets to inform me that, for corporate SMS services, the multinational Orange has a Spanish-only policy here. Rachid continues to ask if there is no alternative, to which Naeem reacts by smiling at his naïveté as an early ICT navigator in the Catalan network society.

Thus, for non-elite transnational migrants who do not belong to any global think-tanks, it is the Spanish language which is always present, not English, both for business and ICT services delivery and for intercultural customer–agent interactions with the Spanish telecommunications market.

Catalan

Catalan has gained a position in the language hierarchy of the Spanish private business sector, as attested, too, in the practices of companies working in the language industry, as well as by multinationals with a strong presence in Catalonia (Alarcón, 2007; Solé et al., 2005). However, at the time of the fieldwork this language was always relegated to a place after Spanish, indicating that its presence and use in the Spanish ICT world is by no means normalised (Plataforma per la Llengua i Consum Català, 2011; Solé Camardons & Torrijos, 2011: 107). Only eight out of the 30 mobile phone operators that I analysed offered Catalan in their call centres (see Table 2.2), and these were the three multinationals (Movistar, Orange and Vodafone), and five smaller companies (including Pepephone and Bankinter Móvil), three of which were based in Catalonia and in the Basque country (RACC Mòbil, Vueling Móvil and Eroski). The three multinationals included Catalan in their customer services partly as a reaction to the strong public criticism that was mobilised by powerful Catalan consumer agencies (FOBSIC & Plataforma per la Llengua, 2007); the last three companies mentioned projected their corporate image as serving the users of minority languages in the Spanish state (RACC Mòbil, for example, presented itself as a 'Catalan' company).

Orange and Vodafone, though, did not offer their official websites in Catalan, and the vast majority of start-up and ICT operators did not offer *any* customer services in Catalan,[35] thus infringing the customers' language rights under Llei 1/1998 on language policy (see Generalitat de Catalunya, 2006), between 2007 and 2009. At the time of writing, Orange still infringes a newer policy (Llei 22/2010; Statute of the Consumer, passed in January 2011), which states that customers in Catalonia have the right to access services in Catalan (written and otherwise) and that companies have an obligation to offer them.

Tellingly, in 2009 *none* of the seven migrant-oriented operators offered customer services in Catalan, except for Lebara, whose call centre workers kindly found an agent who could help me in this language, but on her own initiative and off the record. The institutional ideology behind this corporate distribution of linguistic resources is that migrants would (or should) enter the technological era basically in and through Spanish only (or English, in some cases). In fact, the call centre agents whom I talked to were puzzled when I asked whether they offered customer services to migrants in Catalan. They did not seem to think of that language as belonging to their workplace institution – one of the call centre agents even laughed at the idea of a transnational citizen using this language.

Allochthonous minority languages

In this institutionalised linguistic hierarchy, migrant allochthonous languages are either included in a tokenistic fashion or simply not included at all in the Spanish telecommunications sector's repertoire. When I contacted the 30 mobile phone operators, I could not find any customer services in any of the migrants' languages, not even among the seven migrant-oriented operators, except for two noticeable exceptions: Modern Standard Arabic and Romanian. These two languages, however, were provided occasionally and as an *ad hoc* exception to sort out communication problems on the spot, in customer services provision, or in advertising campaigns. For example, the migrant-oriented operator Happy Móvil offered Arabic only at the weekends after 4 pm, when their Arabic-speaking agent was available. Similarly, Lebara found that one of its many English-speaking agents could help customers in Urdu, but unofficially and on her own initiative, and so only during her work shift.

One remarkable exception to this lack of genuinely multilingual customer services was the most recently launched migrant-oriented operator, Digi Mobil (established in Spain in 2009), whose call centre exclusively consisted of Romanian 'native speakers'. This demonstrates that 'ethnic' MVNOs are revising their management of non-elite multilingualisms (but only recently) in order to meet their new clientele's needs.

In general, though, the migrants' allochthonous languages could be found, inconsistently, only in a few (largely written) uninformative marketing phrases. Moreover, these were generally majority nation-state languages, rather than the codes more often used by migrants, for instance Urdu rather than Panjabi, or Modern Standard Arabic instead of Tamazight, which is the language that many migrants from the Maghreb area in fact speak in Catalonia (Hernández-Carr, 2011: 98). Besides, such phrases are normally issued via automatic translation, which, unsurprisingly, at times renders the services non-accessible, as illustrated on the packaging shown in Figure 2.3, which is a presumably multilingual (but unrealistic) motto employed by the migrant-oriented company MundiMóvil.

Figure 2.3 illustrates the tokenistic multilingual initiatives and the haphazard treatment of the languages of migration by the Spanish telecommunications sector, where the packaging for the company's SIM card includes the Italian, English, Mandarin Chinese, Romanian, Russian, Modern Greek, Urdu, Hindi and French languages. This, though, was evidently done using automatic word-for-word translation, with poor results; for instance, the English version reads 'Calling yours never was so easy'.

Figure 2.3 The telecommunications sector's unrealistic management of linguistic diversity. The MundiMóvil packaging presents the phrase 'The world in your hands' in 10 different languages. *Source*: http://www.mundimovil.es

In order to investigate the reasons provided by the telecommunications sector for this lack of resources concerning the languages of migrants and their 'backgrounding' approach to them, I interviewed the CEO of one of the most important Catalonia-based international ICT ventures. This leading company offers translation via mobile phones in a variety of languages, and it mostly targets migrant-oriented ICT companies or organisations working with or for transnational populations (in fact, it was featured in the news as a 'business opportunity for operators that specialise in "immigrants"'). At the time of the interview, this CEO's enterprise was developing an automatic Spanish–Arabic translator for Movistar, which, in the end, did not get commercialised, as explained in Excerpt 2.5.

Excerpt 2.5. Real multilingual services are 'not worth the investment'
@Location: 29 April 2009. Company headquarters. Vallès Occidental, Barcelona.
@Bck: The CEO of a leading ICT venture tells the researcher (RES) about his experience in trying to assemble a Spanish–Arabic translator for Movistar, which in the end was scrapped.

01 *CEO: *bueno <què passa> [?] que traduir a castellà # en aquest cas era castellà àrab àrab castellà.*
   %tra: well <what's the matter> [?] that translating into Castilian # in this case it was Castilian Arabic Arabic Castilian.
02 *RES: *aha.*
→ 03 *CEO: *és difícil # llavors es pensaven que el nostre traductor ho faria perfecte i clar no ho fa perfecte perquè és un traductor automàtic i llavores <li> [/] li van passar precisament a una persona de: una empresa <de> [/] de telecomunicacions al Marroc.*
   %tra: it's difficult # then they thought that our translator would do the job perfectly and of course it doesn't do it perfectly because it is an automatic translator and so then they passed <it> [/] it on precisely to a person fro:m a telecommunications company <in> [/] in Morocco.
04 *RES: *sí.*
   %tra: yes.
→ 05 *CEO: *i clar no s'ajustava ben bé a l'origen real de les frases i dius +"/.*
   %tra: and of course it didn't adapt well to the real origin of the sentences and you say +"/.
06 *CEO: *+" clar lògic.*
   %tra: +" of course logical.
→ 07 *CEO: *per això <hauríem de> [/] hauríem de fer això que estem fent ara,, <no> [?] aportar un producte específic adaptat a aquelles necessitats llavors això sí que es pot garantir la qualitat però així si vols un traductor genèric per resoldre dubtes tràmits per exemple burocràtics pels immigrants hem de muntar un traductor específic per allò.*
   %tra: to get this <we would have to> [/] we would have to do what we are doing right now,, <right> [?] to provide a specific product which is well suited to those needs and then quality can indeed be guaranteed but if you want a generic translator to answer bureaucratic questions procedures for instance for immigrants we need to create a specific translator for that.
08 *RES: *clar.*
   %tra: of course.
→ 09 *CEO: *llavors funcionarà molt bé # no t'hauràs de: limitar a una manera de dir les frases o així sinó serà lliure funcionarà bé per allò però clar has d'invertir uns recursos -, va quedar allà no vam insistir més perquè -, <i> [/] i ells no insisteixen # és a dir jo crec que això lo que demostra és <la> [/] la falta de necessitat real d'això,, <no> [?].*

%tra: then it will work really well # you won't have to: limit it to one way of saying the sentences or whatnot rather it will be free it will work well for that but then look you have to invest some resources -, we left it there we didn't persist any longer because -, <and> [/] and they don't persist # that's I think that what this demonstrates is <the> [/] the lack of a real need for this,, <right> [¿].
10 *RES: clar.
%tra: sure.
→ 11 *CEO: o la falta de preocupació pels operadors per aquest tipus de gent pues potser perquè per aquest tipus de negoci pel volum que genera és poc.
%tra: or the lack of concern about these sort of people by operators may be because for this type of business the turnover it generates is small.

The first trial of the Spanish–Arabic automatic translator, as the CEO explains, was rejected by a telecommunications enterprise in Morocco (line 3), on the grounds that the Arabic version was faulty and non-realistic (line 5). He admits that a generic translator will not fulfil the specific needs of migrants, such as dealing with administrative procedures (line 7), and adds that this would require a flexible 'free human quality translator' (as automatic translation agents call it) that may specifically suit the migrants' real communicative needs in particular contexts. He states that his company has the technological resources to design it, but points out to the fact that this made-to-measure translator would increase the multinational's costs significantly. Three possible explanations are provided for the fact that the first trial did not succeed and that negotiations with Movistar ended in deadlock. The first is that this multinational does not have a real need for it in order to keep targeting Arabic-using migrant customers (line 9), which implies that the multinational has so far been able to capitalise on this specific migrant niche through Spanish or English, using only generic translators. The second explanation that the CEO provides is that Movistar is simply not concerned about this particular customer group. This may seem contradictory because the multinational is indeed investing resources in trying to figure out how to include Arabic in its repertoire, and because it claims to offer Arabic in its call centres for a potentially large number of clients (this was the last language in the hierarchy offered to me by a Movistar agent in one of their call centres). The last (and, to me, most plausible) explanation is that the multinational has estimated that the turnover that the Arabic-using clientele generates is not worth the investment required for anything more sophisticated than a generic Spanish–Arabic translator (line 11).

This interview helped me understand that the management of linguistic diversity in the telecommunications world is uniquely constrained and organised within the world of linguistic engineering, which is broadly defined as the use of the knowledge of languages 'to develop intelligent information processing systems that are able to recognise, understand and reproduce human language in its various forms' (UNESCO, 2009). It includes fields as disparate as telecommunications, engineering, computing and information systems, mathematics, microelectronics, interface design, image-to-text encryption, automatic translation, speech technology, voice recognition and digital speech, among many others (notably, though, it does not include disciplines related to social sciences). What these fields share is that they tend to see languages as ciphers or as combinations of codes. In fact, within the realms of human language technology the term 'language' is basically employed for naming programming systems, such as 'Java Language', 'Extensible Markup Language' and the like. Besides, the management of linguistic diversity is normally understood as the *programming, synchronisation* and *direct translation* of purely functional, denotational codes, that is, as a technical skill.

Therefore, it should be noted that the developers of multilingual services within the ICT world share a deeply rooted belief that their job and their professional contribution to ICT-mediated communication is *external* to the more social aspects surrounding technology. This is illustrated, for example, by the Movistar consultant's statement with which I opened this chapter: 'Obviously technology has nothing to do with languages; it is the same for everyone'. In setting up new systems and new interfaces, their 'communication problems' are understood as technological, not social, to be dealt with via computer innovation.

For example, the CEO quoted in Excerpt 2.5 told me that the main 'obstacle' to automatic translation of Asian languages such as Mandarin Chinese on mobile phones was that these languages employ sinography (i.e. Chinese characters, or logograms). The 'solution' that the company provided was purely technical and consisted of encoding characters in the widely used UTF-8 system, which allows for the use of many different alphabets on computer. And yet, by 2009 this innovative method had become marginal to their business, since, as he acknowledged, multilingual migrants who were supposed to be the end-users of these services found it impractical and demanded the support of a *social infrastructure* grounded upon their actual daily communicative realities. Therefore, despite the linguistic engineers' genuine efforts to offer multilingual services in allochthonous codes, the treatment of language as a technical skill leads to the non-accomplishment of their entrepreneurial objectives, because the applications that they provide do not meet client demands. These require the support of free or inexpensive in-group *face-to-face human interaction* (for example, the support of their immediate migrant social networks), as the CEO clarified later in our interview when he stated that 'for an immigrant here then perhaps it [our mobile phone translation system] is not necessary, right? And he says "then why do I have to spend half a euro on translating something," right? "I'll just ask anyone around".'[36]

Overall, the data that I have presented on the linguistic ideologies and on the actual public language practices of the Spanish telecommunications sector demonstrate that the sector's multilingual services, at least for the migrant segment of the market, are simply a commercial façade, that is, a *linguistic fetish* (Kelly-Holmes, 2005) which excludes those clients who are non-schooled or non-literate in dominant Western alphabets (basically, the Roman script), those who do not have access to Spanish (or perhaps English or another of the European nation-state majority languages) and those who are non-numerate or unfamiliarised with the prevailing alphanumeric system and the semiotic language nested in the northern hemisphere (Cameron, 2000; Chipchase, 2008; Kress & van Leeuwen, 2006 [1996]).

Therefore, the Spanish telecommunications sector's management of linguistic diversity, instead of shrinking present-day digital divides, imposes, like the nation-state, a marginalising linguistic regime which incapacitates many migrants, and thereby brings about the *new global illiteracies* of the 21st century. Resistance to this linguistic regime is described in the following chapter, where I analyse how it gets mobilised at the margins of the two internationalised technopolitical bodies that currently govern migrants, in the alternative, subversive multilingual spaces of the *locutorios*.

3 *Locutorios* as Challengers to Established Political-Economic Orders and Sociolinguistic Regimes

> *When my boss exploits me, who listens?*
> *Without papers there is no work and without work there are no papers,*
> *the spiral makes me a stranger and I got caught in Spain's net.*
> *What can I do? I breathe hopelessness,*
> *scraping out a living in a dark hideout because my salary is not enough,*
> *I go to the locutorio daily, this is my duty,*
> *to get to know how my people are, what they will have to eat.*[37]
> Nach, 2005 (from the song 'Tierra prometida', on Ars Magna – Miradas)

In this chapter, I explore the ways in which different migrant populations have articulated a series of pragmatic subversive practices against their structural marginalisation in order to informally establish, in a bottom-up manner, a unique 21st-century institution of mundane resistance at the margins of the hostile citizenship regimes and governmentality practices of the internationalised Spanish nation-state and the Spanish telecommunications sector, in *locutorios*.

In the first section, I set out the particular socioeconomic context in which these migrant-operated call shops emerged, and I describe, in detail, how, when, where, why and exactly under what circumstances they took roots as empowering grassroots meeting points, consolidated, somehow unexpectedly in Catalonia, as truly transnational, self-governed spaces of migration 'with an unrivalled force in Europe' (Ros, cited in BBC Mundo, 2008).

In the second part of this chapter, I focus on the daily workings of *locutorios* in order to show how documented and undocumented migrants draw on their transnational social networking capital and manage to gain a degree of both individual and collective social agency, which is necessary to challenge and indeed overcome the institutional legal, economic and, above all, communication barriers that I have analysed in the previous chapter. More specifically, I show that *locutorios* today allow for migrants' self-provision of vital resources (like food, legal advice and shelter) and of communication technology (from SIM cards and cheaper international phone calls to truly multilingual money transfer services) in their host societies.

I conclude by arguing that it is ultimately the overworked *locutorio* employees who, by mobilising their impressive translocal repertoires, end up being the real but silenced and unrecognised multilingual translators and cultural mediators in charge of informally articulating the migrants' transnational survival.

From Autochthonous Local Businesses to Alternative Institutions of Transnational Survival

The term *locutorio*, first recorded in 1580, comes from the Latin word *loquax*, which means 'talker' (Corominas, 2008: 364). Although it was occasionally employed to name the few first shops selling long-distance calls in the cities of the 1980s, it was not until the 1990s that the word started to systematically make reference to international telephone service centres in Spain (*El País*, 1981, 1990).

The first *locutorios* were owned by the company Telefónica, which at the time had a monopoly on international voice telephony. In 1997, with the liberalisation of the Spanish telecommunications sector (Ley 12/1997 de Liberalización de las Telecomunicaciones), newer Spanish businesses specialising in long-distance calls were allowed to set up their own *locutorios*. These enterprises emerged in direct competition with the former monopoly, on being established in partnership with international companies which provided them with the latest technological equipment and with substantial foreign capital and investment, whereby they were able to offer much cheaper prices to their customers (*El País*, 1997).[38]

With the opening up of the market, the estimated number of *locutorios* in 1999 reached 800 in the Spanish state. Around 500 of these were located in Madrid City, and half of them belonged to Vic Telehome, Telefónica's biggest competitor, which by 2001 had accumulated the capital to establish 375 more *locutorio* businesses, selling an average of 15 million international call minutes a month. As its founder and then president Alejandro Loring stated in an interview, the company was successful because it capitalised on the emergence of a niche market, transnational migrant customers: 'The *locutorio* phenomenon is tied to immigration. The boom started in the year

1999 and it has burst now'[39] (*El País*, 2000). In 2002, Comytel, a Valencian telecommunications wholesaler of international telephone services, led by Juan Peiró, took over from Vic Telehome and started offering still more competitive prices for long-distance calls, in partnership with well known North American multinationals (like Cable & Wireless and Global Crossing). It kept investing in *locutorios* and provided a modern infrastructure that was cheap to rent, in order to attract 'foreign' clients in Spain (Consumer, 2004).[40]

Thus, *locutorios* emerged as local ICT businesses which belonged to the Spanish telecommunications sector's elite, and which were formally designed and run in a top-down fashion by individual Spanish (or Spanish-nationalised) entrepreneurs with the financial support of big foreign companies that first decided to capitalise on a 'foreign' clientele, *not* by business-minded migrant entrepreneurs – yet.

During the first decade of the 21st century, though, after the main legalisation campaigns of undocumented foreign residents that took place between 2000 and 2005, the socioeconomic conditions were ripe for self-employed migrants to start establishing their own *locutorio* businesses. This was an attractive option, because it was relatively quick and easy to do so in terms of the bureaucracy and business management. The Catalan administration called *locutorios* a simple *activitat innòcua* (literally, an 'innocuous activity'), which meant that these particular 'ethnic' businesses could be set up by asking the local municipality for a licence, filing for a transfer of ownership, or applying for a refurbishment of an existing property.[41] The first migrant entrepreneurs who set up a *locutorio* chose to ask for a transfer of business ownership (popularly known as *traspassos*), for which they could use the existing technical report and so avoid the need to ask for a new licence. Another alternative that also took roots among new migrant entrepreneurs was to directly sub-let an existing small ICT shop (Moreras, 2007).

During the early 2000s, with the neoliberal restructuring of the Catalan small business sector, many local shops formerly run by Catalans in the metropolitan area of Barcelona became available to migrant entrepreneurs. The former gave over their groceries *en masse* to the latter due to a lack of generational relief and to the increased competition from supermarkets (Serra del Pozo, 2008; Solé *et al.*, 2007). Thus, it was a favourable time for migrant entrepreneurs to find property in which to set up their *locutorios*. Very soon, they forged a place for themselves and filled in that niche market, revitalising such emptied spaces and quickly incorporating within the Catalan local economy (Arjona Garrido & Checa Olmos, 2006).

In most cases, the migrants' take-over occurred without the need to acquire a bank loan, particularly if the infrastructure (the computers, the telephones, the desks and so on) was already available, or if the new owners could count on some transnational income or, alternatively, on informal loans, rotating savings or credit associations arranged via their own social networks, like the *kommitti*, organised among people of Pakistani origin (Parella Rubio, 2005; Solé *et al.*, 2007). Therefore, by the time of the fieldwork, *locutorios* had already become what in the commercial jargon is known as 'circuit' enterprises (Portes *et al.*, 2001), that is, truly transnational ventures managing a *global circulation* of clients, workers, bosses and technological consumer goods (Íñiguez-Rueda *et al.*, 2012; Peñaranda Cólera, 2005, 2011).

The way in which these migrant entrepreneurs capitalised on the *locutorio* market was new, since they employed a series of managerial conditions and business regimes which aimed at self-sufficiency, with less dependence on the Spanish telecommunications sector, starting what I see as the *migrantisation* (or the migrants' gradual appropriation and transformation) of such ventures. While they hired some basic technological equipment from the formal ICT market, multinationals intervened only at the margins, as the new migrant entrepreneurs later recirculated their ICT products and services among themselves through their own self-organised transnational groups, doing business largely in informal networks, *outside* the formal circuits of commerce (Pecóud, 2000). In the neighbourhood of El Paso, for instance, most *locutorio* owners rented their landlines from Movistar, but then employed a program specifically designed for managing public multi-cabin calls (Locutorio Tellink Locutax Plus), which was distributed at first by the Spanish Tellink Sistemas de Comunicación S.L. but later resold from migrant hands to migrant hands. Likewise, the prepaid call cards sold in that area were brought from Barcelona City to the neighbourhood of El Paso without going through the usual circuits, by a Pakistani wholesaler who purchased them from another distributor that was actually based in London.

This flexibilisation or *informalisation* of the *locutorio* business was possible in part because the migrant entrepreneurs could impose a series of exploitative workplace practices upon their (co-national) migrant employees, who, with few other employment opportunities, became cheap labour, without any institutional protection. For the entrepreneurs, this migrant workforce represented a small investment that turned into a big profit, for two main reasons. Firstly, they established a series of money-saving tactics whereby they paid their workers no more than an average of €800 a month for their 12-hour shifts each day, including weekends, at the time of the fieldwork. Secondly, the unique multilingual capitals of these overworked *locutorio* employees were an immediate guarantee of success for the owners because of the customer services they could offer. Such genuinely multilingual migrants, unlike the Spanish telecommunications sector, could then immediately understand the migrant clients' communicative behaviours and technological needs.

The public reports available indicate that by 2007 there were already 747 *locutorios* in Madrid City, mostly run by individual foreign residents (Ramírez, 2007: 15). A year later, in 2008, this number had increased to more than 2,500, according to the Immigration Council of the Spanish government (BBC Mundo, 2008). In Spain as a whole, the number of *locutorios* increased from about 800 in 1999 to more than 25,000 in 2008, according to the Spanish Association of *Locutorio* and Cybercafés, ALYCE (Players4Players, 2008).

Similarly, there is not a great deal of information on the emergence and consolidation of *locutorio* businesses in Catalonia but the few studies available indicate that they became part and parcel of the Catalan urban landscape in the late 1990s, in Barcelona City, in the crowded urban areas of Santa Coloma de Gramenet and L'Hospitalet de Llobregat, and in non-metropolitan localities such as Salt and Banyoles, in the province of Girona, north-east of Barcelona (Íñiguez-Rueda *et al.*, 2012; Peñaranda Cólera, 2005; Roca i Albert *et al.*, 2009; Ros & Boso, 2010; Serra del Pozo *et al.*, 2003).[42]

With the aim of providing more detailed information on the emergence of *locutorio* shops in the urban areas of Catalonia, I here present the case of the county of Vallès Occidental (see Figure 3.1), adjacent to the Barcelonès zone, as an illustration of the rapid growth *locutorio* businesses experienced through the first decade of the 21st century.

According to the official *locutorio* registration lists provided to me in March 2009 by the 23 town halls of the region, the first *locutorio* was registered in 1998. There were only five *locutorios* in the entire county until 2001, when documented migrant entrepreneurs opened 16 new *locutorios* all at once, mostly in the larger towns. The greatest *locutorio* growth occurred in 2004, with 42 new registrations that year. In 2008 there was a second peak, with 32 new *locutorio* registrations, mostly in localities where there had been few *locutorios* or none. Thus, in a span of approximately seven years (between January 2001 and March 2009) the number of *locutorios* in the region increased from 21 to 212, as shown in Figure 3.1.

Figure 3.1, which links the number of *locutorio* businesses with the percentages of foreign residents in 2001 and in 2008 provided by Idescat (2001, 2010a), shows that the increase of *locutorio* registrations went hand in hand with an increase in the percentages of foreign residents, which explains why the areas in darker shades have a higher concentration of *locutorio* shops. In both maps, the localities which welcomed one *locutorio* business had registered at least 500 migrants; the localities with more than one *locutorio* had registered at least 1,500 migrants; and the areas where the number of

64 Migrant Communication Enterprises

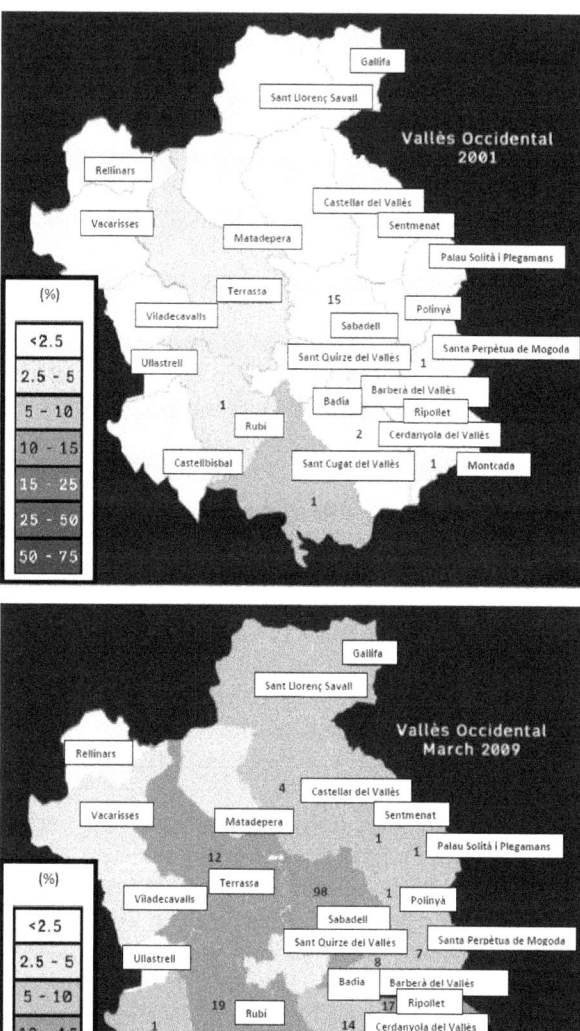

Figure 3.1 Numbers of *locutorios* registered in Vallès Occidental, January 2001 and March 2009. Shading indicates percentage of foreign residents over total population. *Source*: Idescat (2001, 2010a), and official records of the 23 town halls in Vallès Occidental. Data assembled by author. Maps designed by Neus Sabaté i Barrieras

Table 3.1 Percentage of foreign population and number of *locutorios* by town, Vallès Occidental, 2001

Town	Total population (n)	Foreign population (n)	Foreign population (%)	Locutorios (n)
Terrassa	174,756	5,466	3.13	0
Sabadell	185,170	3,178	1.72	15
Rubí	60,303	2,078	3.45	1
Sant Cugat del Vallès	55,825	3,257	5.83	1
Cerdanyola del Vallès	53,481	1,165	2.18	2
Ripollet	30,548	635	2.08	–
Montcada i Reixac	28,714	509	1.77	1
Sta. Perpètua de Mogoda	19.235	357	1.86	1
Barberà del Vallès	26,741	294	1.10	–
Castellar del Vallès	18,136	250	1.38	–
Palau Solità i Plegamans	11,419	175	1.53	–
Castellbisbal	8,696	206	2.37	–
Badia del Vallès	15,032	48	0.32	–
Sant Quirze del Vallès	13,259	188	1.42	–
Sentmenat	5,820	41	0.72	–
Polinyà	4,767	52	1.09	–
Matadepera	7,197	74	1.03	–
Viladecavalls	6,035	72	1.19	–
Sant Llorenç Savall	2,053	19	0.93	–
Vacarisses	2,926	47	1.61	–
Ullastrell	1,159	23	1.98	–
Gallifa	162	7	4.32	–
Rellinars	410	<4	–	–
Vallès Occidental	731,844	18,143	2.48	21
Barcelonès	2,105,302	92,847	4.41%	
Catalonia	6,361,365	257,320	4.05%	

Source: Idescat (2001) and *locutorio* registrations in the 23 town councils of Vallès Occidental until 2001. Data assembled by author

foreign residents was higher than 3,000 had registered 10 *locutorios* or more, as detailed in Table 3.1 (for the year 2001) and Table 3.2 (for the year 2008).

In the March 2009 map we can also see that *locutorio* businesses tend to be located in towns where already in 2001 there were more than 250 registered migrants, that is, in towns with a longer history of in-migration which, in turn, experienced the highest increase in migrant registrations

Table 3.2 Percentage of foreign population and number of locutorios by town, Vallès Occidental, 2008

Town	Total population (n)	Foreign population (n)	Foreign population (%)	Locutorios (n)
Terrassa	206,245	27,918	13.54	12
Sabadell	203,969	24,367	11.95	98
Rubí	71,927	9,814	13.64	19
Sant Cugat del Vallès	76,274	9,793	12.84	12
Cerdanyola del Vallès	58,493	5,808	9.93	14
Ripollet	36,255	3,949	10.89	17
Montcada i Reixac	32,750	3,556	10.86	17
Sta. Perpètua de Mogoda	24,325	2,399	9.86	7
Barberà del Vallès	30,271	1,883	6.22	8
Castellar del Vallès	22,626	1,233	5.45	4
Palau Solità i Plegamans	13,916	944	6.78	1
Castellbisbal	11,795	835	7.08	1
Badia del Vallès	13,829	820	5.93	0
Sant Quirze del Vallès	18,225	806	4.42	0
Sentmenat	7,633	517	6.77	1
Polinyà	7,403	507	6.85	1
Matadepera	8,460	222	2.62	0
Viladecavalls	7,170	189	2.64	0
Sant Llorenç Savall	2,357	161	6.83	0
Vacarisses	5,787	192	3.32	0
Ullastrell	1,761	36	2.04	0
Gallifa	213	16	7.51	0
Rellinars	685	11	1.61	0
Vallès Occidental	862,369	95,976	11.13	212
Barcelonès	2,235,578	381,308	17.06%	
Catalonia	7,364,078	1,103,790	14.99%	

Source: Idescat (2010a) and *locutorio* registrations in the 23 town councils of Vallès Occidental until March 2009. Data assembled by author

over the seven years that I analysed. Finally, both maps also show that *locutorios* tend to be located in (post-)industrial urban areas. Thus, rural localities in the north and west of Vallès Occidental, with no direct connections to Barcelona, had or have few *locutorio* businesses.

These data on the *locutorios* of Vallès Occidental are also meant to illustrate the success of these entrepreneurial initiatives, for the *locutorios* almost

Table 3.3 Average net income generated at the *locutorio* of El Paso (2007–09)

Basic services	Income
Computers/internet	€25–30 a day
Cabins	€85.99 (minimum) to €234.31 (maximum) a day
Top-ups	€4025 a month
International phone cards	€580 a month
Money transfers	€400 a month

Source: Receipts, bills and invoices registered by the main computer program of the *locutorio* in El Paso, collected on a systematic basis during the time of the fieldwork, averaged and assembled by author

displaced the long-distance call shops that belonged to the formal Spanish telecommunications sector, including those with direct links to Telefónica. The *locutorio* of El Paso certainly generated a substantial net income. At the time of the fieldwork, it welcomed between 61 and 156 customers a day, who made between 79 and 211 cabin calls (also a day) and who spent about €6,000 a month on telecommunications there, which allowed the Pakistani *locutorio* owner (*not* the worker) to earn a monthly salary of approximately €4,500. In fact, his enterprise was so profitable that he was soon able to open three other 'ethnic' businesses in the nearby, even under the shadow of the economic downturn in Catalonia, in the year 2008 (for the financial details on the *locutorio* of El Paso, see Table 3.3).

The entrepreneurial success of *locutorios* brought about a phase of aggressive competition on the part of the formal Spanish telecommunications sector. Multinationals like Movistar, for instance, had to lower the discount plans of their international calls by 45% in order to be able to compete with the prices offered in these migrant-operated call shops (Consumer, 2004). Simultaneously, they also started to present themselves as alternative 'grassroots' businesses, through the emulation of the *locutorios'* informal shop front-stage, to the extent that they also employed the label *locutorio* to advertise their products and services, like the *locutori mòbil d'Orange* ('Orange's mobile *locutorio*') or the *mini-locutorio de Movistar* ('Movistar's mini-*locutorio*), presented in Figure 3.2.

This commercial rivalry led to a series of lawsuits promoted by a group of Spanish entrepreneurs whose 'autochthonous' internet centres were being challenged by the migrants' commercial initiatives. The elite Asociación de Locutorios y Cíbers Españoles (ALYCE, Association of Spanish *Locutorios* and Cybercafés), for instance, claimed that rivalling migrant *locutorio* owners did not pay VAT, apparently evading €19 million a year (Players4Players, 2008).

Tarjeta Internacional
el locutori mòbil d'Orange

Figure 3.2 Multinationals' competition techniques against 'ethnic' *locutorio* businesses. The '*Locutori mòbil*' (Orange) (left) and the '*mini-locutorio*' (Movistar) (right). Pictures taken by author. Vallès Occidental, 28 November 2007

Likewise, the Associació d'Empresaris de Locutoris de Catalunya (AELCOM, Association of *Locutorio* Owners in Catalonia), founded in 2004 by a Spanish entrepreneur whose own recreational internet café was being threatened by newer Pakistani-run *locutorios* around Barcelona, filed similar accusations against 180 such establishments, claiming that they had no licence and that most of them bought cheaper telephone connection minutes from the Canary Islands in an illegal way (Ràdio Hospitalet, 2004).

Pressure upon *locutorio* owners at the time of their expansion came not only from the Spanish telecommunications sector but also from the bureaucratic machinery of the Catalan government. The local municipalities started a period of strict regulation and close surveillance of these 'ethnic' call shops. Catalan town halls, at first initially considered the *locutorios* a type of temporary or marginal business and, consequently, classified them very unsystematically, coming under different administrative sections of the town council. They also used at least 10 different labels in Catalan for their entry on the official registers.[43] The underestimation of the *locutorios'* potential growth was such that in 2009 I could only find a single entry under the label '*locutorio*' in the official Chambers of Commerce Registries of both Barcelona and Sabadell (similar problems are reported in Feliu *et al.*, 2012).

This, to a certain extent, attests to the Catalan administration's unease in managing and integrating *locutorios* into the workings of the municipalities. It also speaks of the authorities' unwillingness to see migrant businesses as an integral part of the Catalan local economy. In fact, since 2002, many town halls in Vallès Occidental have reacted to the largely unexpected consolidation of *locutorios* by establishing a series of municipal ordinances (at times accompanied by door-by-door inspections) restricting their specific geographical location, their opening and closing times, their internal layout and their display of written information.[44] Some of the measures taken during this period of surveillance included: the obligation to have two toilets, separate for men and women, and a waiting room with chairs; a prohibition on accommodating more than 30 clients at once (the standard maximum capacity is 50); the obligation to locate new *locutorios* at least 250 metres from each other and at least 50 metres away from public schools, health centres and buildings employed for religious purposes, and in streets with at least a two-metre-wide pavement; and, finally, a prohibition on the public display of any type of written information outside the *locutorio* (Ajuntament de Terrassa, 2009; BOP, 2002, 2005, 2006). As a clerk in one of these town councils put it to me (in Catalan), 'we are establishing a set of conditions which are practically impossible for them to fulfil.'[45]

I argue that the ideological rationale behind these measures shows that the Catalan administration presented *locutorios* as marginal 'ghettos', with an 'ideology of mistrust' (Codó, 2008: 90) that publicly institutionalised their stigmatisation, stereotypation and criminalisation, at times with racist overtones. This is shown in the following quotation, taken from a municipal ordinance regulating the establishment of *locutorios* in the town of Rubí, which links these businesses with a decrease in the quality of life of the citizens inhabiting that locality:

> The success and proliferation of these services together with the absence of a specific technical legislation cause an increased nuisance and parallel activities that may result in a decrease in the quality of citizens' lives.[46] (Town council of Rubí; Ajuntament de Rubí, 2005)

In a similar manner, the mass media were also quick to foster the idea that 'parallel activities' which threatened security and the well-being of citizens were being conducted in and around *locutorio* shops. The press, for instance, has linked *locutorios* with migrant-organised terrorism, drug dealing, money laundering and prostitution ever since these call shops started colonising the Catalan urban landscape (see, by way of example, *Catalunya Press*, 2012; *Diario de León*, 2004; *El 9 Nou*, 2012; *El Punt*, 2010; *Qué*, 2009).

And yet, despite this recent history of fierce competition from the Spanish telecommunications sector, of regimentation and prosecution by the Catalan local administration and of criminalisation by the press, *locutorios* have consolidated their position as the migrants' preferred alternative spaces of communication, as explained in detail in the following section.

Locutorios as Successful Transnational Points of 'Meetingness' and Mundane Resistance Practices

Locutorios have become the migrants' favoured place to access not only telephony services for both local and international communication but also internet services, regardless of whether these ICT resources are available at home or not, for they do not only act as technology shops but also as key transnational meeting points, at which to gain access to crucial transnational social networks (Feliu *et al.*, 2012: 200; Molins Pueyo, 2006: 5; Ros & Boso, 2010: 144).

In the fieldwork, the centrality of *locutorios* to the migrants' transnational lives and prospects was apparent in the narratives of most informants. Merche, an undocumented babysitter and cleaner in her forties born in the Dominican Republic, for example, kept joking about the fact that she used the *locutorio* so frequently that she felt that she practically lived there, as illustrated in Excerpt 3.1 (lines 2 and 4).

Excerpt 3.1. Locutorios *as central to the migratory experience (I)*

@Location: 22 August 2008. *Locutorio* in El Paso. Vallès Occidental.
@Bck: The researcher (RES) asks Merche (MER) whether she uses the locutorio frequently.

```
 01  *RES:  viene a menudo¿
     %tra:  do you come frequently¿
→02  *MER:  sí bastante # ya te digo casi vivo aquí [=! laugh].
     %tra:  yes quite a lot # let me tell you I almost live here [=! laugh].
 03  *RES:  <sí> [¿] <casi vive aquí> [¿][=! laugh].
     %tra:  <really> [¿] <you almost live here> [¿][=! laugh].
→04  *MER:  ya te digo casi vivo aquí [=! laugh].
     %tra:  I'm telling you I almost live here [=! laugh].
```

Similarly, in Excerpt 3.2 Aboubacar ('Abou'), a 36-year-old scrapyard worker from Guinea, also explained that he used the *locutorio* on a daily basis (line 3), prompted by Abde's teasing him, in a quick overlap and with a humorous, emphatic tone, about his *locutorio* visits several times a day (in line 2).

Excerpt 3.2. Locutorios as central to the migratory experience (II)

@Location: 1 September 2008. Bar near the *locutorio* in El Paso. Vallès Occidental.
@Bck: Teased by Abde (ABD), Abou (ABO) explains to the researcher (RES) that he visits the locutorio on a daily basis.

```
   01 *RES:   y así que <cuántas veces vas al locutorio al día> [¿].
      %tra:   and so <how many times a day do you go to the locutorio>
              [¿].
→  02 *ABD:   +^ éste cada cinco minutos entra baja entra sube entra
              [=! laugh].
      %tra:   +^ this [guy] every five minutes goes in comes back goes
              in comes over goes in [=! laugh].
→  03 *ABO:   <cada:> [/] cada día.
      %tra:   <every:> [/] every day.
```

One of the reasons why *locutorios* have emerged as an integral part of everyday life for migrants like Merche and Abou is the fact that, apart from being communication technology ventures in tune with the needs of mobile citizens, they have also become a key space where crucial transnational local networks are negotiated, established, maintained and actualised, as explained by Roca i Albert *et al.* (2009: np):

> *Locutorios*, the result of the entrepreneurial initiative of recent immigrant citizens, [...] distribute relevant information that is important for everyday living and for [migrant] incorporation into the city. The *locutorio* is more than just a centre of private contacts [...]. It is a place where they can run into acquaintances and where information of a personal or collective nature can be exchanged.[47]

This is further illustrated in Excerpt 3.3, where Abde, from Morocco, talks about the *locutorio* as a local 'place of meetingness' (Urry, 2007: 254).

Excerpt 3.3. Locutorios as transnational spaces of local meetingness

@Location: 28 August 2008. Bar near the *locutorio* in El Paso. Vallès Occidental.
@Bck: Abde (ABD) talks to the researcher (RES) about the local transnational networks of people in El Paso who interact and socialise in their comings and goings at the *locutorio*, the local market and the bar.

01 *RES: has encontrado amigos allí¿
   %tra: have you made friends there¿
   %com: 'Allí/there' refers to the locutorio.
→ 02 *ABD: hombre como está junto el locutorio con el barrio sí que xxx amigos porque la misma gente.
   %tra: well since the locutorio is alongside the neighbourhood yes xxx friends because the same people.
03 *RES: <sí> [¿] <la gente de aquí también> [¿].
   %tra: <really> [¿] <the people from here too> [¿].
→ 04 *ABD: es la misma gente.
   %tra: it's the same people.
05 *RES: cómo los conocías¿
   %tra: how did you meet them¿
→ 06 *ABD: vienes encontrar gente al bar de locutorio al mercadillo siempre al bar pero aquí al barrio.
   %tra: you come run into people at the bar from locutorio to street market always in the bar but here in the neighbourhood.

In Excerpt 3.3, Abde explains that he started meeting people with whom to share transnational experiences in the *locutorio* because this has become an integral part of the neighbourhood (line 2), and because the *locutorio* clients are the same friends and acquaintances who also visit the bar round the corner (line 4), another space of transnational socialisation. When I ask him how he met his *paisanos* or 'compatriots' (Abde used this label to address his Moroccan friends and other Muslim acquaintances at the time of the fieldwork), he states that in this neighbourhood one gets to meet people in the *locutorio*, the market and the bar (line 6). This suggests that networking points like *locutorios* are spaces where to attain increased *social networking capital* and gain access to many local contacts as well as to groups of transnational mobile people of different backgrounds (see also Feliu *et al.*, 2012; Íñiguez-Rueda *et al.*, 2012; Peñaranda Cólera, 2005, 2011).

Thus, today *locutorios* are central local meeting points of safe 'mooring' (Hannam *et al.*, 2006: 2), that is, protective or umbrella institutions at which to articulate a series of resistance practices against social dislocation, fixity and entanglement, and against collective institutionalised marginalisation and social inequality practices, for the following reasons. Firstly, *locutorios* subvert the dataveillance systems of the Spanish nation-state by selling SIM cards to the undocumented, providing technological equipment to those who would otherwise be, in theory, disconnected. In the *locutorio* of El Paso, for example, these call cards were legally purchased in packs of

10 by the Pakistani owner, who used his corporate identification number (his Certificado de Identificación Fiscal, or CIF) to register and later resell phone numbers under his venture's commercial name, skirting Ley 25/2007 on compulsory data storage (discussed in Chapter 2). Besides, *locutorios* also offer unlocking services (the so-called *móviles libres* or *móviles liberados*), which allow for the use of more than one operator on a single handset and with a single telephone number without having to notify any change to the company from which the first SIM card was purchased, that is, without having to provide any personal identification. (This service cost between €30 and €120, depending on the type of handset.)

Secondly, *locutorios*, unlike multinationals and regular ICT companies, are specifically geared towards the economic conditions and the budget constraints of transnational migrant clients, because they offer cheaper prices for international calls (though, notably, not for local ones). A one-minute call to Pakistan, for instance, cost 10 cents at a *locutorio*, whereas the cheapest discount plan offered by the multinational Movistar in July 2011 cost 15 cents a minute. Besides, these ventures also offer, informally, personalised in-group-regulated credit-giving services when telephone invoices cannot be paid, subverting the strict commercial norms that characterise the managerial regimes of the formal private business world, as exemplified in Snapshot 3.1, taken from my field notes from 30 June 2008.

Snapshot 3.1. Informal loans at the locutorio: *A way to access ICT on a limited budget*

A young man with a Latin American Spanish accent that I cannot quite identify approaches the desk slowly, and Naeem, automatically, presses the key to issue his cabin receipt. It is €13, a fairly large sum compared to the average for this *locutorio*, which is the 30-cent bill resulting from the typically brief one-minute local call. Head bowed, he tells Naeem that that was not what he had planned to spend, that the call took unexpectedly longer, *'lo siento mucho'* ('I'm terribly sorry'), apologising in a very soft voice. He then takes his wallet and empties it all on the desk. He also searches his pocket without success. After that, he takes his temporary residence permit out of his wallet, puts it onto the desk next to the few coins, and tells Naeem *'toma'* ('here you are'), promising to be right back. Naeem panics at the idea of having someone else's proof of citizenship in his hands, '¡No, no, no, no!' he shouts, and tells the young client that the deal is to be back with the money in two days, at the latest.

Internet services, though cheaper than in 'regular' recreational cybercafés, cost 25 cents per quarter of an hour. In the metropolitan area of Barcelona, migrants have the option to access workstations with online access for free, in municipal libraries or telecentres, which were set up by the Catalan government in several towns and cities in order to reduce any 'digital divide' (BOP, 2008: 119). However, these spaces require the provision of personal details and compulsory registration at the town hall, which explains why many migrants are deterred from visiting them, most obviously the undocumented, who fear being reported. Ros and Boso (2010: 144) reported that only 3% of migrants access ICT services in such civic spaces, and only 2% in community centres. There was a municipal library and a free-access workstation in the neighbourhood of El Paso, but none of the informants whom I observed made use of them.

Thirdly, the *locutorios* uniquely offer a wide range of ICT resources in one single space (telephone services, international call cards, internet access, money transfers, fax and photocopy services, and the like). This is especially advantageous for migrants, because they need more than one service when communicating transnationally. For example, during their *locutorio* visits they normally use the phone cabins immediately after arranging some remittances (the migrants' ICT-mediated practices are presented in Chapter 4).

Fourthly, unlike the banks and other financial institutions offering remittance services, *locutorios* adapt not only to the migrants' economic conditions but also to their busy timetables and to their transnational activities across different time zones. This is explained by roommates Yousaf, a 42-year-old electrician from Pakistan, and Shabbir, the 41-year-old construction worker from Kashmir, in Excerpt 3.4.

Excerpt 3.4. A social infrastructure adapted to migrants' transnational schedules
@Location: 22 August 2008. Bar near a *locutorio* in El Paso. Vallès Occidental.
@Bck: Yousaf (YOU) and Shabbir (SHA) explain to the researcher (RES) that they usually send remittances via *locutorios* because Spanish banks do not take foreign time zones into account.

01 *RES: y qué le gusta más Ria o el banco?
%tra: and what do you like most Ria or the bank?
%com: Ria is an electronic money transfer company available at locutorios.
02 *YOU: hombre Ria es un servicio para: rápido.
%tra: well Ria is a service fo:r express.

```
     03  *RES:  vale.
         %tra:  ok.
→    04  *YOU:  cuando tú <tienes> [//] necesitas urgente vale.
         %tra:  when you <have> [//] need express it's ok.
     05  *RES:  para urgente.
         %tra:  for express.
     06  *YOU:  y normalmente con el banco también me gusta.
         %tra:  and normally with the bank I also like it.
     07  *RES:  vale.
         %tra:  ok.
→    08  *YOU:  pero peor en el banco a veces dicen +"/.
         %tra:  but it's worse in the bank sometimes they say +"/.
→    09  *YOU:  +" ah no hombre hoy no llega o hoy no tiene tiempo y ya
                está ya he plegao y +...
         %tra:  +" ah no man it isn't arriving today or today [I] have no
                time and that's it I'm off and +...
     10  *RES:  en Pakistán¿
         %tra:  in Pakistan¿
     11  *YOU:  en Pakistán también aquí no no no # hace falta mucho
                <(¿)> [>] pero en mi país también.
         %tra:  in Pakistan as well here no no no # there is still a need
                <(¿)> [>] but in my country as well.
→    12  *SHA:  <de horario> [<].
         %tra:  <of opening times> [<].
     13  *RES:  sí.
         %tra:  yes.
→    14  *SHA:  porque tiene abierto a la ocho y hasta la una en Pakistán # el
                banco está cerrado pero el locutorio no.
         %tra:  because it's open at eight and until one in Pakistan # the
                bank is closed but the locutorio is not.
     15  *RES:  claro.
         %tra:  sure.
→    16  *SHA:  el locutorio no locutorio hasta las diez once están abiertos.
         %tra:  the locutorio is not locutorio until ten eleven they are open.
     17  *RES:  claro.
         %tra:  sure.
→    18  *SHA:  tú llegas y puedes.
         %tra:  you get there and you can.
     19  *RES:  +^ puedes mandar.
         %tra:  +^ you can send.
```

When I ask Yousaf whether he prefers to send remittances via regular banking or via the money transfer company that he had previously mentioned at the beginning of our interview (Ria), he states that Ria, which is presented as being more effective than banks (line 8), is useful for express, urgent services (line 4), for two reasons. Firstly, express services may take longer to arrive through banks (line 9). Secondly, he agrees with Shabbir – who, in an overlap, helps clarify Yousaf's explanation about the lack of longer opening hours at banks – in that with formal financial institutions there is a problem with scheduling (line 12). Banks do not take foreign time zones into account and, unlike *locutorios*, they are closed when Pakistani banks are open for people to pick up their remittances (line 14). By contrast, *locutorios* are open until later (line 16) – many operate until midnight, including weekends, in Catalonia. Thus, they are the easiest and quickest way to send remittances worldwide (line 18).

Apart from subverting the sorts of legal regimes and economic constraints that I have just outlined, another reason for migrants to choose to patronise *locutorios* is the fact that they also provide (and are provided with), circulate and redistribute basic material resources which may have nothing to do with technology but which are required for transnational survival. For instance, inside *locutorios* homeless people may be given access to bathroom facilities and to temporary shelter when it rains. Besides, migrant networks of support distribute food, water and cigarettes and, moreover, offer access to informal jobs, to rooms for rent and to transportation, as illustrated in Snapshot 3.2 (field notes from 22 and 23 August 2008).

Snapshot 3.2. The provision of basic resources for transnational survival in locutorios

It is the end of August in El Paso and it starts raining during the night, as if to warn us that the summer break is almost over. Luis, the second generation of a family from southern Spain, has been quick to grab one of the hard-fought-for chairs facing the computer desks. When night falls he starts having his dinner, which he carries in a plastic bag with his other belongings. It consists of canned tuna, a beer, and some cigarettes, which he smokes depending on Naeem's willingness to engage in a chat (he routinely gets Naeem's water and cigarettes without asking, once he's done with his). Around 11 pm a tall black man comes in, wet from feet to toe. He must be a mechanic, for his clothes are covered in a thick coat of grease. Without saying anything, he stops at the entrance, gives Naeem a quick supplicatory look, his dirtied arms wide open, and points his head towards what has become the public toilet of the neighbourhood. *'Vale'* ('ok'), sighs Naeem.

The morning wakes up busier. Merche drops by, half asleep, and, in a rush, leaves a huge box with some shopping, assuming that Naeem will take good care of it. Without hesitation, she also picks up a piece of paper and sticks it upon the *locutorio*'s wall, the real heart of this place, right before leaving. It is a hand-written room-for-rent advertisement (in Spanish), looking for a *chica latina* (a 'Latina' young woman): she and her partner need a roommate (apparently from a similar ethnolinguistic background) with whom to share household expenses. Yousaf honks from the street twice, and his flatmate Shabbir leaves in a rush. They both need to go to Barcelona, for which Yousaf, who doesn't have a valid driving licence, provides the car, and Shabbir, who doesn't have a vehicle but has a Spanish driving licence, becomes the driver.

Locutorios have also become a 'weapon against ignorance' (Luna René *et al.*, 2007: np), since their associated transnational migrant groups gatekeep another resource which deserves attention: key information such as knowledge of the host societies (including informal help in writing CVs, free legal advice and support concerning legalisation procedures, or tips on filling in bureaucratic forms), which, before, migrants had to find in more formal, top-down institutions such as state offices, town halls or pro-migrant NGOs. An example of this is presented in Figure 3.3, which shows a leaflet offering legal services with free consultation, including (in Spanish): 'Family reunification, work permits, temporary residence permits for exceptional circumstances in the event of foreigners rooted in Spain, student permits, self-employment, equivalence of academic credentials, driving licences, revocation of court orders of expulsion' and 'dual citizenship'.

Locutorios also make up for the language gaps left by the Spanish telecommunications sector, which tends to welcome them in Spanish or perhaps in other dominant global or nation-state majority languages such as English, and only very unsystematically in Catalan, with allochthonous codes such as Arabic and Romanian playing a marginal role (see Chapter 2). By contrast, in *locutorios* users can access information on a wide range of ICT products and services in many of their non-elite codes and in non-Roman

- REAGRUPACIÓN FAMILIAR
- PERMISOS DE TRABAJO
- PERMISOS DE RESIDENCIA POR ARRAIGO
- VISADO DE ESTUDIANTE
- AUTÓNOMO
- HOMOLOGACIÓN DE TÍTULOS
- PERMISOS DE CONDUCIR
- REVOCACIÓN ORDEN DE EXPULSIÓN
- DOBLE NACIONALIDAD

93
CONSULTA GRATUITA

Figure 3.3 The migrants' self-distribution of key resources (information). Leaflet offering legal services with free consultation in the *locutorio* of El Paso. Picture taken by author (with selected details removed). Vallès Occidental, 30 September 2008

alphabets, thereby subverting the linguistic orders and literacy regimes of the Spanish private business sector. An example of this was provided by Sheema, a 31-year-old bar owner born in Pakistan, who managed to get information about a Movistar discount plan from a friend who translated and typed the important pieces of information about that particular promotion from Spanish into Urdu, on his own initiative, rendering it accessible to Urdu readers, as presented in Figure 3.4. The *locutorio* worker then photocopied and distributed copies to the literate members of the Pakistani local network, resisting and subverting the established linguistic landscape of that particular multinational company.

Finally, the *locutorios*' commercial success can also be explained by their unique informal multilingual customer services in many of the migrants' languages, provided by the *locutorio* workers, who, with both their command of some allochthonous codes and their knowledge of (and experience with) the prevailing communicative regimes of the Catalan network society, end up doing the multilingual job of the Spanish telecommunications sector, thus becoming the real language mediators and articulators of these alternative institutions of migration and mundane resistance, as argued below.

Figure 3.4 Services provision in non-recognised migrant languages. Leaflet with a Movistar discount plan in Urdu, translated from Spanish on a migrant's initiative. *Locutorio* in El Paso. Picture taken by author (with selected details removed), 4 September 2008

The crucial provision of technoliteracy: Informal language work

Normally, the migrants' allochthonous languages are highly stigmatised, devalued and excluded from the Catalan local economy. In the market niche of the *locutorios*, though, they have become the necessary but largely uncosted and unseen *worktool* required in the provision of face-to-face customer services in an effectual manner. Thus, in these discursive spaces the employees' multilingual repertoires have become uncompensated but key commercial assets or profitable 'commodities' (Heller, 2003; Tan & Rubdy, 2008) with which different groups of transnational clients may be more directly approached, understood and targeted. Consequently, most *locutorio* workers, who are in theory employed only as ICT shop assistants, have entered the Catalan marketplace by virtue of their distinct multilingual capitals (which includes experience with the host societies' languages), becoming the 'language workers' who make these small businesses not only work but prosper (Boutet, 2006, 2008, 2012; Duchêne, 2011; Heller, 2010b; Heller & Boutet, 2006), in this case from the silenced margins. Among the growing pool of migrants who have learnt that they are a 'linguistic marker of exploitability' (Alarcón & Heyman, 2013: 2), many are willing to mobilise this unique human capital at zero cost for their employers, under hard (at times exploitative) working conditions because they need the job in order to keep their temporary residence permit, or simply as a way to avoid unemployment (see also Chapter 6).

Locutorio workers have thus become the ultimate articulators of these alternative institutions of migration because, in their sometimes monotonous work, crucially *articulated in and through language*, they end up acting as *ad hoc* interpreters (Baynham & Lobanga Masing, 2000; Bührig & Meyer, 2004), that is, as linguistic 'brokers' (Martin-Jones & Jones, 2000), informally mediating between the Spanish telecommunications sector and those migrant clients who are not socialised into the dominant Western literacy-based cultures. Therefore, it is mostly these language workers who, in reality, turn these ventures into grassroots 'schools of technology' (BBC Mundo, 2008) and carry out the 'technological socialisation' of migrants (Feliu *et al.*, 2012). They provide disconnected migrants with the technoliteracy capitals required to overcome established linguistic barriers in the information and communication age, as I shall now illustrate.

The main tasks that *locutorio* employees accomplish routinely include: the translation of text messages from Spanish into many other languages for non-literate clients; the dialling of telephone numbers for non-numerate migrants who bring their contact information on a piece of paper; and even the filling in of administrative forms or official documents for those who are not familiar with the bureaucratic culture in Catalonia and, more particularly, those who lack the knowledge to deal with it successfully. An instance of how this informal language mediation works is provided in Snapshot 3.3 (from 28 August 2008), which explains how the *locutorio* worker in El Paso begins to socialise one of his non-literate clients from Romania into the technoliteracy world of the global ICT system, here presented in and through the written Spanish language only.

Snapshot 3.3. The informal provision of technoliteracy capitals by locutorio *workers*

Archid needs to check his official labour situation to make sure that he is finally registered with social security, but he can't figure out how to do this online (he was given a number and was told to check it on the internet). The *locutorio* worker, his *amigo* 'Paki', without hesitation approaches his computer and Googles *'informe leboral. es,'* which the program self-corrects into *'quiso decir* [you meant] *"informe de vida laboral"* [work life].' They both click on the first link and get to the www.seg-social.es website, becoming post-modern learners and expert users of the search tool, which is here employed as a way to overcome non-literacy within the Western-orchestrated Spanish-only telecommunications system.

Another example of the ways in which *locutorio* employees help migrants navigate Western modern bureaucratic practices is provided in Figure 3.5, which shows an online money transfer receipt which the *locutorio* worker in El Paso, Naeem, filled in for a non-literate client born in Senegal who found the form in Spanish inaccessible.

Money transfers such as the one presented in Figure 3.5 are normally ultimately issued by the formal remittance facilities of the Spanish telecommunications sector, but these are sub-let by a *locutorio* owner and actually

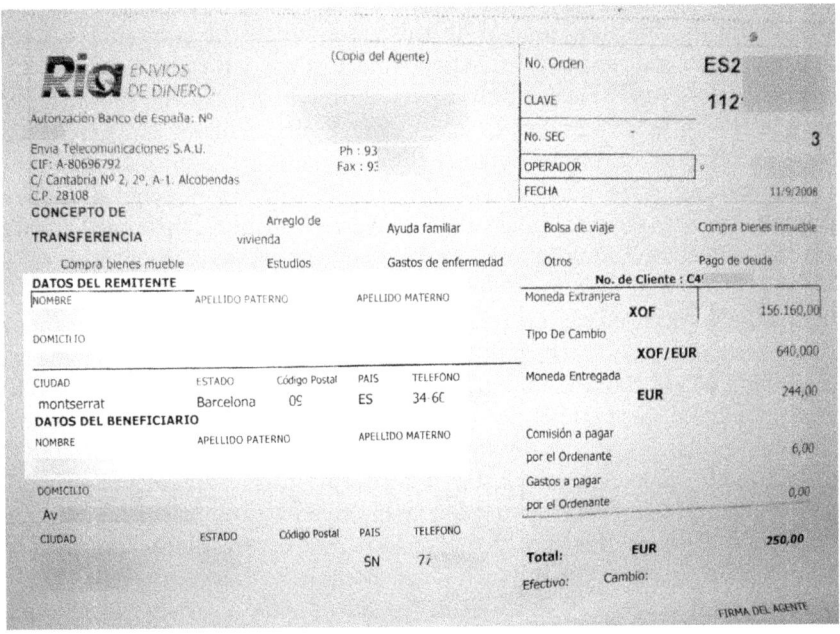

Figure 3.5 The intercultural mediation practices of *locutorio* workers. Scan of a receipt of a successful money transfer written in Spanish (with selected details removed). *Locutorio* in El Paso, 11 September 2008. (Highlighted details discussed in text)

put into use by a migrant *locutorio* worker. The online information required is not always readily available to many migrants, because it in most cases is issued in Spanish only. Naeem, the *locutorio* worker who has no problems communicating in Spanish, also encounters some intercultural misunderstandings and some information gaps when filling in such applications for his clients. However, he has developed the pragmatic, tacit knowledge required to overcome them quickly. In the receipt (Figure 3.5), the terms *apellido paterno* (father's surname), *ciudad* (city), *estado* (state) and *código postal* (postcode) have been misunderstood and have been filled in 'wrongly' (highlighted area in Figure 3.5). Under the father's surname, for instance, Naeem wrote a common Senegalese first name – in Pakistan the family surname is rarely used (Moyer, 2011). Also, he has written the name of the street (Montserrat) under 'city', and under 'state' he has provided the name of the province (Barcelona) rather than 'Spain', although under *país* ('country') this is properly indicated (ES), perhaps because in Pakistan a 'state' roughly corresponds to a Spanish province. Finally, the postcode is wrong (it should begin with 08, not with 09). Naeem, though, quickly learnt that he needs to double-check only the address of the payee, the country of destination (Senegal, in this case, which the computer system automatically codes as SN) and the amount of money to be sent. In fact, the Senegalese payer needed only a password, which I assume that he later gave to the payee at a branch office in a major city in Senegal, also selected by the electronic system by default.

Banks and state post offices in Catalonia offer this type of money transfer service as well, normally via Western Union, the official state agency, whose only help offered to migrant customers is the lending of a 'Western Union multilingual translator'. This translator consists of a compilation of plasticised sample forms that are a word-for-word translation of the Standard Spanish form into dominant nation-state majority languages (including Modern Standard Arabic, Bulgarian, Mandarin Chinese, Dutch, English, French, German, Hungarian, Italian, Polish, Portuguese, Romanian, Russian and Urdu). In the towns of Vallès Occidental that I investigated, I encountered only one such post office offering this 'multilingual' service, and this had, unsurprisingly, not been employed by any of the informants I followed; Ros and Boso (2010: 144) report that approximately only 1% of migrants use post offices for such aims. Because in banks and post offices transnational citizens have to fill in the documents by themselves, those who lack the procedural knowledge to deal with Spanish bureaucratic forms find this system completely useless. By contrast, in *locutorios* they readily find a multilingual person who, despite many difficulties, can act as the *mediating link* between the institutional communicative regimes of banks or other multinationals and the clients' multilingual world, empowering migrants with the capacity to become functionally technoliterate in the ICT realm, via a largely successful customer services provision that better understands the clients' communicative needs.

In short, *locutorios* provide migrant populations with a migrant-operated *social infrastructure* whereby they can individually and collectively articulate the fight not only against the legal barriers and the economic constraints but also against the linguistic marginalisation posed by the two advantaged technopolitical governmentality blocks of the 21st century, the nation-state and the telecommunications sector. This is the main reason why, partly thanks to the uncredited but pivotal role of their language workers, *locutorios* have become a truly alternative institution of transnational survival, which allows for migrants' incorporation into the Catalan network society, through their own means and in their own ways.

4 The Self-Provision of Technological Capital in *Locutorios*: A Diversity of ICT-Mediated Networking Practices

> *The modification of communication processes by the interaction between [...] social practice and a new range of communication technologies constitutes indeed a profound social transformation.*
> (Castells *et al.*, 2007b: 246)

In this chapter, I approach *locutorios* as windows into the largely unexplored migrants' ICT-mediated social networking strategies and actual transnational communication practices in Catalan urban areas. I claim that what migrants do with the wide range of technological resources that they now have available in these institutions is rather subversive, in the sense that their actual communication behaviours and non-conventional calling routines overtly challenge the unrealistic business plans specially designed for them (outside *locutorios*) by the Spanish telecommunications sector. Thus, I show that migrants have appropriated the products and services offered by the private ICT business sector in order to develop their own alternative consumption tactics, clearly engaging with the global consumerist cultures, but in migrant-regulated call shops which allow them to skirt many aspects of the sector's regimes.

I first analyse the ways in which migrants have gained some degree of autonomy from multinationals and from telephone companies when deciding how, when, where, with whom and for how much to establish and to maintain ICT-mediated transnational contacts. In the second section, I present an array of practices which shows that migrants have developed innovative ways to redistribute their technology resources both across diverse migrant groups and within their own networks, cleverly managing their own self-connection into the Catalan network society, in a social mobilisation which is unprecedented. In the last section, I argue that the diverse calling schemes that may be observed inside *locutorios* also demonstrate that, against the widespread (and perhaps ethnocentric) idea of displacement, uprootedness, distress and emotional weakness, migrant populations have largely organised their kinship roles and their responsibilities within transnational family units in a very effectual manner, by *doing* everyday parenthood and by actively practising virtual friendship; in other words, by fully normalising their transnational living via ICT, regardless of physical distances.

Individual Mobility Projects and Subversive Communication Technology Tactics

From the privileged space of the *locutorio* of El Paso, I studied the migrants' varied selective use of public telephone cabins, mobile phones, text messages and internet services, and recorded the ways in which they frequently combined and simultaneously adopted all these technology devices in nuanced manners and for different communication purposes, in Catalan urban areas.

Firstly, I observed that, contrary to common thought, public landline telephony is crucial for migrant populations to maintain local contacts within transnational networks of support in their host societies. This goes against the Spanish telecommunications sector's entrepreneurial schemes, which more often than not present the migrants' communication practices as always 'international' and unidirectional, from the migrant's host societies to his or her 'country of origin'. In fact, most of the products or services offered to migrants by multinationals are basically discount plans for international calls to a single 'foreign country', that is, packages for communicating with people abroad.[48] This is shown in more detail in Figure 4.1, which shows an advertisement issued by the company Lebara. It presents two short biographies about a Moroccan breadwinner leaving his home alone in order to support his family back in Morocco, and about a Pakistani young man migrating temporarily to Spain to save money and pay for his wedding 'back home'. These brief transnational life stories, which basically emphasise north-to-south contact across transnational family units, read (in Spanish): 'Mounir from Morocco does not lose contact with his daughter, who lives in Tangier with her mother', and 'Tashkir from Pakistan does not lose contact with his girlfriend, whom he will marry at the end of the year'.

And yet, an analysis of the actual use of telephone calls in the cabins of the *locutorio* of El Paso reveals that the most frequently dialled telephone numbers by migrant *locutorio* users are local mobiles, *not* international

86 Migrant Communication Enterprises

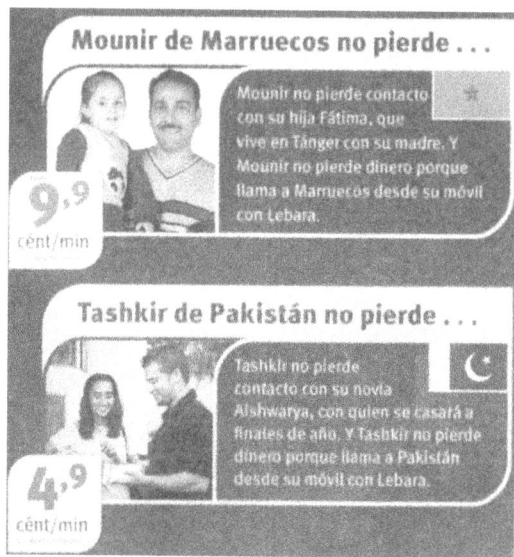

Figure 4.1 The market's discursive and visual representations of ICT-mediated 'transnational communication' (*Metro*, 2008b: 17)

landline or mobile phone numbers, suggesting that, for these transnational populations, communication technology is important not only to keep international ties but also to establish, actualise and maintain *local* transnational contacts in Catalonia. This is exemplified in Table 4.1, which presents the number of cabin call clients and the number of calls per user, registered on eight different days (chosen from the first and the last weeks of the months, including weekends and week days); the main call destination on each particular day (overwhelmingly local mobiles); the longest and the shortest calls to local mobiles (indicating their minimum and maximum duration, in minutes) and the main destinations of landline calls and of calls to mobile phones other than to local mobiles.

Table 4.1 also shows that the calls to local mobiles were usually short, lasting from a few milliseconds (i.e. 00:00 min) to less than eight minutes. They tended to be of an instrumental nature. This is seen in the field notes that I gathered while I was collecting the cabin call receipts. They included extremely brief conversations giving information of a pragmatic kind related to the organisation of everyday transnational life, like: 'Estoy en casa de la yaya. Llámame al móvil que se me va a cortar' ('I'm at granny's. Call me on

Table 4.1 Migrants' use of *locutorio* telephone cabins

	Mon.	Sun.	Mon.	Tue.	Wed.	Thu.	Fri.	Mon.
	30/6/08	7/7/08	28/7/08	29/7/08	30/7/08	31/7/08	1/8/08	1/9/08
Number of users (*n*)	68	61	71	98	74	98	156	138
Number of calls	79	81	89	144	107	114	211	178
Main call destination of the day	Local mobile	Local mobile	Local mobile	Local mobile	Local mobile	Local mobile	Local mobile	Local mobile
Min. duration of calls to local mobiles (min.)	00:22	00:02	00:01	00:00	00:02	00:00	00:02	00:02
Max. duration of calls to local mobiles (min.)	03:16	06:00	02:27	07:45	06:17	05:27	04:51	07:23
Main call destination for landlines	Local	Poland	Pakistan	Pakistan	Pakistan	Local	Pakistan	Cameroon
Main international call destination for mobiles	Senegal	Morocco	Morocco	Ecuador	Morocco	Romania	Romania	Senegal

Source: Cabin call receipts issued automatically by the computer program of the *locutorio* of El Paso, collected in a systematic manner during the time of the fieldwork. Data taken from Sabaté i Dalmau (2012c: 201–202).

the mobile, I'm cutting it short'); and 'Oye, que ya te lo he mandado, ¿Vale?' ('Hey listen, I've already sent it to you, okay?').

The shortest calls did not actually involve any talk (clients hung up after making sure that they had heard the first dialling tone) and functioned as what is called the 'beeping' or the 'flashing' system (Overa, 2008: 48). This consists of the use of missed intentional calls, employed, for example, to let a friend know that one is ready to meet, or that one has managed to catch the 17:30 pm train as planned, without having to engage in a conversation and saving a few cents as well – the clients who made frequent use of missed calls came into the *locutorio* with 30 cents ready in hand, which was the cost of a one-minute call to a local mobile (*locutorio* cabins use a per-minute charging system). This is a widespread subversive practice (particularly for undocumented migrants), in the sense that the consumers who engage in it manage to transmit information without having a contract with any company and without having to show or to prove legal citizenship to the *locutorio* workers, thus skirting the gaze of the nation-state. Besides, through the 'missed call' system migrants do not have to pay for any connection fees. They can also control their budget closely when using a *locutorio* telephone cabin, because they can see the minutes and seconds that they have used up on a small screen placed next to it.

A second observation I made in El Paso is that migrants frequently organise their transnational ICT-mediated communications by *strategising* with the many different technologies and communication services that are readily available in *locutorios*. Shabbir, the construction worker from Kashmir, for example, employs cabin services as well as his home landline and his two mobile phones, simultaneously using prepaid international and local call cards in ingenious, nuanced ways, as illustrated in Excerpt 4.1.

Excerpt 4.1 Strategising with different ICT services

@Location: 10 August 2008. Street near a *locutorio* in El Paso. Vallès Occidental.
@Bck: Shabbir (SHA) explains to the researcher (RES) how he changes mode, place and payment method of communication depending on the person he wants or needs to talk to, by strategising with many technology devices in order to establish a transnational communication that best suits his particular needs and his budget.

```
01  *RES:   y entonces y: <utilizas el locutorio a veces> [?].
    %tra:   and so a:nd <do you use the locutorio sometimes> [?].
    [...]
```

→ 02 *SHA: cuando yo quería hablar con mi hermana # yo tengo una hermana.
   %tra: when I wanted to talk to my sister # I have a sister.
   03 *RES: ah vale.
   %tra: ah ok.
   04 *SHA: y: habla mi hermano mi madre padre todos querías hablar cuando llegamos y con mi mujer y dos hijos también.
   %tra: a:nd my brother my mother father talk you [they] all wanted to talk when we [I] arrived and with my wife and my two children as well.
   05 *RES: claro.
   %tra: of course.
→ 06 *SHA: y cuando quería hablar con toda familia sí que puede usar una tarjeta pero cuando sólo quería hablar con mi mujer cinco diez minutos usando el locutorio.
   %tra: and when I wanted to talk with the entire family yes it is then possible to use a card but when I wanted to talk to my wife five or ten minutes using the locutorio.
   07 *RES: el locutorio.
   %tra: the locutorio.
   08 *SHA: sí.
   %tra: yes.
   09 *RES: vale <es más económico el móvil o el locutorio> [¿].
   %tra: ok <[what] is more economical the mobile or the locutorio> [¿].
→ 10 *SHA: más económico es eh tarjeta.
   %tra: more economical is eh card.
   11 *RES: la tarjeta¿
   %tra: the card¿
   12 *SHA: sí.
   %tra: yes.
   13 *RES: esta tarjeta <vale> [/] vale.
   %tra: this card <ok> [/] ok.
   %act: Shabbir shows the researcher a prepaid call card for landlines.
   14 *SHA: de cinco [euros] +...
   %tra: of five [euros] +...
→ 15 *RES: +^ a través del fijo,, <no> [¿].
   %tra: +^ through the landline,, <right> [¿].
→ 16 *SHA: sí.
   %tra: yes.

When Shabbir wants to make quick international phone calls (5–10 minutes, normally to talk to his sister or to his wife, who, he says, demand very frequent calls), he chooses a public space with cheap international phone call cabins: the *locutorio* (lines 2 and 6). When he has to speak for longer to members of his extended transnational family with whom he talks only every now and then, in Kashmir and in Pakistan, he prefers to use prepaid call cards (line 10), which are apparently even cheaper, and which he buys at the *locutorio*, too. He uses his telephone landline (lines 15 and 16), which is shared with an uncle of his in order to divide the expenses, to talk privately from the quietness and intimacy of his flat. Thus Shabbir chooses his spaces of communication (the *locutorio*, the home, or the street for mobile phone calls) as well as the mode, type and payment method of communication, according to the speaker (or speakers). He does so by strategising with the devices, the companies' prices and the different types of relationship that he maintains with peers, workmates and relatives, demonstrating that transnational migrants adapt communication technology to their own needs, rather than submit to the logic of the technology (Castells et al., 2004: 239).

Apart from cabin calls and prepaid phone cards, Shabbir also uses two mobile phones very regularly. He chose to have two mobile phones for economic reasons: he could save money by having one handset that worked with a particular rate for international calls, and another one for local calls. Thus, one of Shabbir's mobiles was in theory used only to *receive* international calls (he just had to have some credit on it, in case he did need to contact his family quickly when he was not in the *locutorio*). The other mobile phone, with a flat rate for making mobile phone calls within the Spanish state, was meant to be employed to initiate and to receive only local calls.

This 'double SIM' strategy is very popular among migrants, and its use has become generalised in the neighbourhood of El Paso, where many migrants routinely carry two handsets (alternatively, some use only one handset but swap between two SIM cards manually). Sheema, the Pakistani bar owner, explains how this 'double SIM' strategy works in Excerpt 4.2.

Excerpt 4.2. The 'double SIM' strategy

@ Location: 10 August 2008. Bar near the *locutorio* in El Paso. Vallès Occidental.
@Bck: The researcher (RES) tries to figure out why informants like Sheema (SHE) make very frequent use of two mobile phones in and around the locutorio of El Paso.

01 *RES: cuántos móviles tienes¿
   %tra: how many mobiles do you have¿
→ 02 *SHE: dos.
   %tra: two.
03 *RES: <dos> [¿] <cómo es que tienes dos> [¿].
   %tra: <two> [¿] <how come you've got two> [¿].
→ 04 *SHE: no porque uno aquí la normal uno con tarjeta pero que:: ##
           tengo uno pero uno estropeado # sólo uno.
   %tra: no because one here the normal one with the card but tha:t
           ## I've got one but another [is] broken # only one.
   [...]
05 *RES: y el otro <para qué lo utilizas> [¿] <no lo utilizas> [¿].
   %tra: and the other one <what do you use it for> [¿] <don't you
           use it> [¿].
→ 06 *SHE: <no> [/] no yo tengo línea sí pero <no> [/] no es sólo el de
           la tarjeta.
   %tra: <no> [/] no I have line yes but <no> [/] no it's only the
           one with the [top-up] card.

At the beginning of the interaction, Sheema explains that he has two mobile phones (line 2). When I ask him why this is so, he replies that one is a 'normal' telephone (that is, a top-up mobile) and the other is one he got when he formalised a contract with a multinational (as I discovered later), which he does not use very frequently. Then, perhaps with some hesitation, he self-corrects and tells me, again, that in practice he only has one telephone, not two, because his second mobile seems not to be functioning properly, or to be broken (line 4). When I insist and ask about that apparently unused second mobile, Sheema explains that he does actually carry two connected handsets, but that when he initiates phone calls he utilises only the top-up mobile (line 6), the one employed for local calls, because with this one he can better monitor his budget for telecommunications. Thus, in the end, I understood that, as outlined above, the second handset, with rates for a chosen 'foreign country' which are adapted to a specific foreign time zone, is always connected but used only for international emergency calls (I observed that for planned, long and relaxed international calls he frequently employed *locutorio* cabin calls instead).

The 'double SIM' strategy is yet another subversive technique to access mobile phone services which has many advantages. It allows migrants to choose to be connected to two different places at the same time, but also to choose to be disconnected from one of them when they wish. Besides, the two handsets do not have to be with the same company (in fact, what prevails is to use a Spanish multinational's rate for local calls and an 'ethnic' MVNO operator's offer for international ones), so migrant users can skirt contracts (and, hence, compulsory registration), without getting trapped into the Spanish telecommunications sector's regimes.

In Excerpt 4.3, Vitória, a young woman from Brazil who arrived in El Paso from elsewhere in Spain in 2008, explains that what made her decide to use a 'double SIM' strategy was the fact that, during a mass-media promotional campaign, a multinational gave her a second handset for free, even though she declined to have a contract with that particular company.

Excerpt 4.3. Subverting the multinationals' marketing campaigns

@Location: 25 August 2008. Bar near a *locutorio* in El Paso. Vallès Occidental.
@Bck: Vitória (VIT) explains to the researcher (RES) how she took advantage of a marketing campaign to fulfil her communicative needs and to organise her budget for telecommunications via the 'double SIM' strategy.

```
       01  *RES:  qué compañía utilizas?
            %tra:  what company do you use?
   →   02  *VIT:  tengo un Vodafone y un Movistar.
            %tra:  I have a Vodafone and a Movistar.
       03  *RES:  tienes dos?
            %tra:  you have two?
       04  *VIT:  sí.
            %tra:  yes.
            [...]
   →   05  *VIT:  uno lo compré y otro me lo regalaron.
            %tra:  I bought one and they gave me another one as a gift.
   →   06  *RES:  ah muy bien y <los utilizas los dos> [?].
            %tra:  ah great and <do you use both> [?].
   →   07  *VIT:  sí.
            %tra:  yes.
       08  *RES:  uno para unas cosas y otro por otras?
            %tra:  one for some things and another one for other things?
       09  *VIT:  vale.
            %tra:  ok.
       10  *RES:  uno para aquí?
            %tra:  one for here?
   →   11  *VIT:  no la compañía Vodafone es <mejor> [/] mejor para llamar
                  en Brasil.
```

%tra: no the company Vodafone is <better> [/] better to call Brazil.
12 *RES: ah vale.
%tra: ah ok.
→ 13 *VIT: y la otra es la mejor para aquí.
%tra: and the other is the best one for here.

Vitória uses two top-up mobiles, with Vodafone and Movistar (line 2), though she had intended to have just one. She later changed her mind and decided to take advantage of the fact that a company was giving away handsets for free (line 5) as part of a promotional campaign. She now utilises two mobiles (lines 6 and 7), one for Brazil (line 11) and another one for local calls (line 13). However, she does not make use of the handsets in the ways that Vodafone or Movistar may have expected. Firstly, she has no exclusive contracts with either of them, and she has no contract with either (both her mobiles work with top-ups), so, presumably, the company has not managed to attract her as a permanent, loyal customer through this type of marketing strategy. Secondly, while she participated (and clearly participates) in the telecommunications market's formal circuits of commerce, she also chose what the sector was offering to migrants for her own transnational communicative interests, like Shabbir and Sheema.

The calling tactics described here suggest that migrant ICT users, fully engaged in the consumer culture of the capitalist northern hemisphere, are not simply mere 'passive' consumers, but what in the commercial jargon is actually known as *prosumers* (Rheingold, 2002: 202), that is, 'professional' consumers with the autonomy to interfere in the market via the establishment of their own technology consumption habits, which, I argue, are propelled by initiatives like the establishement of migrant-tailored *locutorio* sites. This is particularly so in the specific types of ICT-mediated communication practices that may be observed within transnational family units, which involve, for example, the informal redistribution and collectivisation of ICT across different family members, as analysed in the following section.

Transnational Family Units and the Collectivisation of ICT

The communicative practices of transnational family units, which are extremely complex and heterogeneous, frequently involve the simultaneous use of different subversive calling 'tricks'. The family-regulated

ICT-mediated practices of informant Juana, a member of the transnational network of migrants in El Paso, are an example.

Juana is a middle-aged woman from Equatorial Guinea who came to Barcelona at the age of 10 with her mother and her brother. In 1999, when her family moved to the town where El Paso sits, Juana decided to settle in London, where she had her second child and started studying nursing. She now travels from London to Catalonia a couple of times a year and stays for a month (I interviewed her during one such stay). She and her family then established a series of communicative routines whereby they found a successful and economically viable way to stay in touch, for which some of the gadgets in the *locutorio* (cheap international and interprovincial calls, internet access, and so on) became decisive, as presented in Excerpt 4.4.

Excerpt 4.4. The complexities of family-regulated transnational communication

@Location: 20 August 2008. *Locutorio* in El Paso. Vallès Occidental.
@Bck: Juana (JUA) describes her family's complex ICT-mediated subversive communicative practices to the researcher (RES).

```
   01  *RES:  usted tiene teléfono móvil¿
       %tra:  do you have a mobile phone¿
   02  *JUA:  sí.
       %tra:  yes.
   03  *RES:  <sí> [¿] y: <cómo lo utiliza es Vodafone # Movistar> [¿].
       %tra:  <yeah> [¿] a:nd <how do you use it is it Vodafone # Movistar> [¿].
→  04  *JUA:  Movistar # <cuando estoy aquí> [/] cuando estoy aquí uso el de mi madre.
       %tra:  Movistar # <when I'm here> [/] when I'm here I use my mother's.
   05  *RES:  ah.
   06  *JUA:  me apodero de él ya ves [=! laugh] y como estoy aquí no: normalmente un mes o así no: eh # es Movistar.
       %tra:  I take it over you see [=! laugh] and since I'm here no: normally for a month or so no: eh # it's Movistar.
   07  *RES:  vale <es de recarga> [¿].
       %tra:  ok <is it a top-up> [¿].
→  08  *JUA:  lo recargo y todo sí.
       %tra:  I top it up and all that yes.
   09  *RES:  bien y entonces <cuándo utiliza el locutorio> [¿].
       %tra:  good and so <when do you use the locutorio> [¿].
```

→ 10 *JUA: cuando tengo que llamar fuera -, como tengo familia en Madrid y a fuera de: +...
%tra: when I have to call abroad -, since I have family in Madrid and away from: +...
11 *RES: +^ ah vale.
%tra: +^ ah ok.
→ 12 *JUA: en distintos sitios de España y si tengo que llamar a Londres también.
%tra: in different places of Spain and if I have to call London too.
13 *RES: es más económico¿
%tra: is it cheaper¿
→ 14 *JUA: claro <desde luego> [!].
%tra: sure <of course> [!].
15 *RES: vale muy bien <y cuando está a Londres para llamar acá> [¿].
%tra: ok great <and when you are in London to call here> [¿].
→ 16 *JUA: eh desde Londres para llamar acá utilizo el teléfono de casa porque tengo un paquete que me entran llamadas a España.
%tra: eh from London to call here I use the landline telephone because I have a package which includes calls to Spain.
[...]
→ 17 *JUA: y ahora mismo estoy con una que se llama Talk Talk que es.
%tra: and right now I'm with one called Talk Talk which is.
18 *RES: +^ Talk Talk¿
19 *JUA: sí.
%tra: yes.
20 *RES: anda! [=! laugh].
%tra: I see! [=! laugh].
21 *JUA: y que está bien también sí.
%tra: and it's ok too yes.
22 *RES: vale y mantiene <cont> [//] <contactos con Guinea> [¿].
%tra: ok and do you keep <cont> [//] <contact with Guinea> [¿].
[...]
→ 23 JUA: <mantengo> [//] tengo contactos de Londres a España y España ya me informa sobre Guinea [=! laugh].
%tra: I <maintain> [//] have contacts from London to Spain and Spain already keeps me posted about Guinea [=! laugh].
24 *RES: muy bien.
%tra: great.

Carmen, Juana's mother, first had a landline telephone, but invoices were not paid and, in the end, it was cut off. Now Juana's mother, brother and sister-in-law all use the same mobile phone, in a prepaid arrangement with a weekly budget that they have stipulated, where calls of a strictly instrumental nature predominate (going to work is the main reason for taking it from the family's shared flat). Juana is in charge of calling them from London once a week, through a package with a fixed rate with the company Talk Talk, which is activated through her British telephone landline (lines 16 and 17) – the cheapest option, she claims, after having tried some offers by the British multinational BT. With her in-laws in London, as well as with her family members who moved to other parts of Spain, she finds it more convenient to use the *locutorio* when she is in El Paso (as stated in lines 10, 12 and, very emphatically, in 14). During the visits to her mother, Juana entitles herself to the use of her family's mobile (line 4), conceptualised as a shared communicative tool. Also, she tops it up when she exceeds the stipulated budget for the week (line 8). She chats with her nieces living in the neighbourhood of El Paso when she is in London, since they are frequently allowed to get a budgeted 15-minute internet connection (25 cents) in the *locutorio*. Finally, it is her mother who, as the head of this transnational family, is in charge of keeping her informed about her two sisters living in Equatorial Guinea (line 23).

This is an organisational arrangement of transnational communication which hinges upon the family unit, and which again consists of strategising with the price, the location and the role of each member within the family (the mother, in this case, being the centre). These complex configurations go beyond established market rationalities in that, for instance, they challenge the widespread assumption that each individual will need, afford or want to use a mobile phone as a strictly personal tool. Thus, the ICT-mediated practices observed among migrants also challenge the market's idea that handsets are personal objects which are not conceived for non-individualistic (family or network) use, and that it is the individual and not a *collectivity* who decides upon when, how and with whom to use it, on his or her own initiative.

In fact, my third observation is that, on many occasions, telephone handsets become, unwillingly or not, a *communal resource* for networking that belongs to, and is regulated by, the transnational family, not individuals, who now adapt and adjust their communicative needs to those of their relatives. Abde, the unemployed construction worker, for instance, provides his family with the SIM cards he purchases at the *locutorio* of El Paso whenever he meets his family in Morocco (he is undocumented and has difficulties buying them elsewhere), as explained in Excerpt 4.5.

Excerpt 4.5. The collectivisation of ICT

@Location: 28 August 2008. Bar near a *locutorio* in El Paso. Vallès Occidental.
@Bck: Abde (ABD) explains to the researcher (RES) how he has taken several handsets to his family in Morocco over the past two years, endowing his relatives with ICT-mediated networking capital.

   01 *RES: tú tienes móvil¿
           %tra: do you have a mobile phone¿
   02 *ABD: hombre móvil sí.
           %tra: well mobile phone yes.
   03 *RES: <sí> [¿] <cuándo te lo compraste o te lo regalaron> [¿].
           %tra: <yeah> [¿] <when did you buy it or they gave it to you> [¿].
→ 04 *ABD: no mi móvil perder cuando voy a Marruecos y volver perder la tarjeta.
           %tra: no my mobile lose when I go to Morocco and come back and lose the card.
           %com: The SIM card.
   05 *RES: en Marruecos¿
           %tra: in Morocco¿
→ 06 *ABD: sí claro yo he dejado dos veces mi tarjeta y por eso no puedo contactar con todo el mundo.
           %tra: yes of course I've left my card twice and that is why I can't contact everybody.
           […]
→ 07 *ABD: cuando voy a mi país mi hermano <ha quitado> [//] <me lo quita> [/] me lo quita.
           %tra: when I go to my country my brother <has taken> [//] <he takes it from me> [/] he takes it from me.
   08 *RES: ah.
   09 *ABD: digo +"/.
           %tra: I say +"/.
→ 10 *ABD: +" toma para ti.
           %tra: +" here you are it's for you.

Abde first explains that he comes back with no SIM cards when he travels to Morocco (line 4). It is not that he loses them, though: he is actually his transnational family's technology distributor there (line 10). More specifically, he is his brother's SIM card provider (line 7), despite the fact that he does not like taking the risk of losing his own contact numbers whenever he visits him (line 6). This is a non-individualistic practice which distributes technology and, therefore, communicative networking capital throughout the (male) members of an extended transnational family. Putting technological devices into communal use becomes a subversive tactic in the sense that this collectivisation is informally organised on the migrants' own initiative, outside the formal circuits of telecommunications, and without contracting with any of the multinationals for plans designed specifically for cross-border calls.

Maintaining Emotional Ties: *Doing* Family From a Distance

Despite the different communicative tactics that allow for a better redistribution of networking capital, the practices established within transnational family units are complex and fraught with difficulties with regard to the renegotiation of each member's roles and responsibilities in the joint endeavour of protecting, caring, raising or schooling loved ones when they inhabit different parts of the globe.

With advertisement campaigns like 'your people at only a cent's distance'[49] (see Figure 4.2), the Spanish telecommunications sector approaches such difficulties in various ways, frequently by presenting transnational family ties or relationships as solely based on late capitalist economic duties and economic dependency.

Multinationals and telephone companies, for instance, tend to talk about ICT-mediated transnational communication as a strong moral obligation

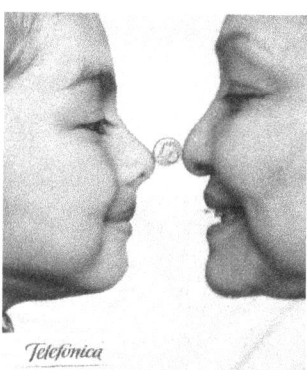

Figure 4.2 The market's construction of economically based transnational family relationships. Telefónica's campaign. Picture taken by author. *Barcelonès*, 22 June 2013

that rests entirely upon the migrating people's shoulders, categorising those who do not frequently call or keep in touch with their relatives 'back home' as irresponsible family carers, or as hypocritical, mean friends, or even, in extreme cases, as *non-persons* (Thurlow & Brown, 2003), with discourses such as: 'To call home you only need to want to' (Movistar leaflet, issued in the Barcelona area between October 2007 and January 2008), or 'A friend is someone with whom talking does not cost a thing' (advertisement by Hits Mobile; ADN, 2008: 15).

On the other hand, the members of transnational family units in migrant-sending societies are at times constructed as pauperised, controlling persons who, exerting long-distance vigilance, hinder the upward socioeconomic prosperity of migrant individuals in their host societies, as if those left behind were an expensive economic burden, with a rhetoric which includes discourses like 'A mobile phone for them, fewer expenses for you' (Movistar leaflet, issued in the Barcelona area between October 2007 and January 2008).[50]

These commercial constructions of late-capitalist family configurations as mostly based on relationships of consumption are also gendered. Migrant men in (Western) host societies tend to be seen and presented as either heroes or betrayers, that is, as making every effort to maintain, protect and maximise transnational incomes and household resources, or else as 'bad sons', 'unconcerned husbands' and disloyal 'womanisers' who space out calls, or authoritarian, tyrannical figures who, by contrast, call too frequently (Moran-Taylor, 2008; Peñaranda Cólera, 2011; Saracho & Spodek, 2008).

Women who migrate without their children tend to be admired as the new masculinised breadwinners who overcome traditional gender roles and patriarchal regimes, but they are also frequently criticised through a deeply rooted argument that delegitimises biological mothers who do not raise their children on their own (or who share mothering tasks with other women). Simultaneously, they are also categorised as self-sacrificing, lonely and emotionally family-dependent women in need of 'reeducation' in terms of motherly habits and routines (Hondagneu-Sotelo & Avila, 1997; Pedone & Gil Araújo, 2008; Uy-Tioco, 2007). This homogenising 'docile' image of transnational women has even led to the publication of some advice on how to handle transnational motherhood on the part of some social educators, with scripted tips for 'weak women' communicating with transnational children which include: 'Don't cry', 'use loving language' and 'avoid silence' (see Rodríguez, 2008: 24–25).

And yet, actual ethnographically grounded studies of current transnational families show that migrating members do not tend to perceive their trajectories in terms of uprootedness, abandonment, complete emotional distress or total loss of control, but rather in terms of *social adjustment* and *family-building continuity*. Thus, contrary to what is assumed and portrayed in the telecommunications world, transnational migrants, men and women, tend to normalise the practice of presence in physical absence, generating ICT-mediated proximities and clearly adapting to a very complex culture of bonds (Diminescu, 2008; Lewin Tapia, 2004; Moran-Taylor, 2008; Peñaranda Cólera, 2011; Saracho & Spodek, 2008). I shall now illustrate this with examples taken from the members of transnational families who talked about their family communication experiences in the *locutorio* of El Paso.

James, a Cuban middle-aged writer and protest poet who arrived in Catalonia in 2008, talks about the difficulties he faces when managing the pressure of feeling forced to maintain constant contact with his 76-year-old mother, via the instant messaging program Messenger (today no longer in use), which leads him to talk about a troubling type of communication which he classifies as 'empty', in Excerpt 4.6.

Excerpt 4.6. 'Empty' communication

@Location: 3 August 2008. *Locutorio* in El Paso. Vallès Occidental.
@Bck: James (JAM) tells the researcher (RES) about the type of transnational communication he maintains with his mother, which he classifies as 'boring' talk that emerges out of routine.

→ 01 *JAM: cuando empieza me corta el rollo porque yo: # <no puede ser> [!] siempre con lo mismo +...
%tra: when she starts it's a pain because I: # <this shouldn't be> [!] it's always the same story +...
%act: looking at the computer while chatting with his mother.
02 *RES: se conecta muy a menudo¿
%tra: do you go online frequently¿
→ 03 *JAM: pues casi todos los días.
%tra: well almost every day.
04 *RES: casi todos los días <con su mamá o también otros familiares> [¿].
%tra: almost every day <with your mum or with other relatives as well> [¿].
→ 05 *JAM: con mi mamá básicamente y alguna amistad.
%tra: with my mum basically and some friends.
06 *RES: vaya <la echa mucho de menos> [¿].
%tra: I see <do you miss her a lot> [¿].

→ 07 *JAM: pues no # me aburre [=! laugh].
   %tra: well no # she bores me [=! laugh].
  08 *RES: le aburre¿
   %tra: she bores you¿
→ 09 *JAM: claro si todos los días está todo el rato tic tic!
   %tra: of course everyday she is all the time tic tic!
   %com: Imitating the sound of the computer keypad when the keys are pressed.
   [...]
  10 *RES: y cuánto rato pasa más o menos cada día¿
   %tra: and how long do you spend more or less every day¿
→ 11 *JAM: diez o quince minutos no más con ella.
   %tra: ten fifteen minutes not more with her.
  12 *RES: +^ con ella.
   %tra: +^ with her.
→ 13 *JAM: claro lo necesario lo mínimo para que ella me cuente sus andanzas.
   %tra: of course the necessary the minimum so that she can tell me all about her adventures.
  14 *RES: sí¿
   %tra: yeah¿
→ 15 *JAM: y que si la nieta tal se cayó y la otra se levantó y aqué:l # <tonterías> [!].
   %tra: and about whether such and such granddaughter fell down and the other one stood up and that o:ne # <silly things> [!].

James explains that the type of transnational contact that he has established with his mother is a sort of 'obligational constraint' (Tilly, 2007: 14) that at times interferes with his work (line 9), which requires the daily use of the internet, since part of his tasks consists of selling his poems online. He emphatically states that when he left Cuba he started experiencing a communicative routine that bores him (lines 1, 7), which is why sometimes he ends up keeping online chats to the minimum (lines 11 and 13). Besides, he comments that he finds his relationship with the past distressful (he left Cuba in 1994 and, after living in many different parts of Spain, he still has some disturbing memories of family problems). He also complains that communication is not always inherently good, and that having the impulse to stay in touch with loved ones on a daily basis at times makes the type and the contents of conversations merely banal, leading to what has come to be known as 'the conversations of yawns' (Lewin Tapia, 2004: 8), 'half-dialogues' (Fortunati, 2002: 51), pseudo-conversations with fillers devoid of interest or feelings, or, simply, repetitive conversations without communication: '*tonterías*' or 'silly things' (line 15).

However, James' case also shows that, despite the 'empty' communication he experiences, he does feel the need to stay in touch with his mother, and he makes it clear that he wants to talk to her on a daily basis (lines 3 and 5) – in fact, I later learnt that he shipped a PC illegally to Cuba and then relied upon a Cuban neighbour to adapt it to the Cuban system in order to be able to chat with her. This goes against the multinationals' simplistic representations of relatives as a 'nuisance'. To James, these conversational situations seem to be part of a normalised management of each other's roles within the intimacy and confidence of the transnational family unit. For many years now, he has been able to exchange small talk and 'good morning' and 'good night' messages with his mother, accomplishing the caring function of checking whether everything is in order 'back home', with a feeling of co-presence.

Another way of modulating one's *autonomy* within a transnational family is to hold the reins in deciding what time is best to establish transnational communication with one's relatives. Shabbir, for example, presents the ways in which he has a say in limiting international phone calls to his wife in Excerpt 4.7. In line 5, he explains how he asked his wife to adapt to his work routines, not the other way around, maintaining a notable degree of social agency as a powerful head of his transnational family unit.

Excerpt 4.7. Migrants' autonomy in establishing transnational contact

@Location: 10 August 2008. Bench near a *locutorio* in El Paso. Vallès Occidental.
@Bck: Shabbir (SHA) explains to the researcher (RES) how his wife in Kashmir adapts to the timetable that best suits his own needs and work schedules when conducting transnational communication.

```
01 *SHA:  mira este es de mío el número de mi mujer.
   %tra:  look this is mine my wife's number.
   %act:  Shabbir shows the researcher a telephone number dialled on
          his mobile.
02 *RES:  ah vale.
   %tra:  ah ok.
03 *SHA:  ahora mismo [imagina que] me llama y yo hablo con ella.
   %tra:  right now [imagine] she calls me and I talk to her.
```

04 *RES: sí.
   %tra: yes.
→ 05 *SHA: él me pregunta que por qué no llamas y yo digo que vale que puedes llamar por la mañana pero hoy no.
   %tra: [s]he asks me why you don't call and I say that ok that you can call in the morning but not today.

Shabbir's case clearly contradicts the market's representation of a docile migrant person being continuously monitored and overwhelmed by the responsibility of ultimately having to sustain transnational contact in order to meet the family demands. He demonstrates that he has the agency to adjust his ICT-mediated contact or his degree of 'social volume' (Baron, 2008) when deciding upon when, how and with whom he talks, at times prioritising his personal timetables and work routines before starting to do kinwork and to provide (or to be provided with) family support.

Merche, the undocumented woman from the Dominican Republic in her forties who is going through her legalisation process, comes to the *locutorio* on a daily basis to make use of the internet and the telephone cabins in order to chat with both her 13-year-old son and her brother, who takes care of him. Her case, presented in Excerpt 4.8, is another example of how migrants manage to modulate the amount of ICT-mediated talk, to feel and *do* family work, and to successfully normalise transnational life in physically distant family relationships.

Excerpt 4.8. The normalisation of transnational family roles

@Location: 22 August 2008. *Locutorio* in El Paso. Vallès Occidental.
@Bck: Merche (MER) explains to the researcher (RES) how she monitors her teenage son's use of ICT and how they both organise their transnational communications successfully, 'doing' family, regardless of the physical distance.

01 *RES: <puede <visitar> [//] <ir de visita> [//] volver> [¿].
   %tra: <can you <visit> [//] <go for a visit> [//] come back> [¿].
→ 02 *MER: ahora sí porque ya estoy regularizándome y ya puedo ir y venir pero ahora hace cuatro años que no los veo a ninguno.
   %tra: I can now already because I'm already going through regularisation [legalisation] and I can already go and come back but I haven't seen any of them for four years now.
   [...]

→ 03 *MER: para mi es una eternidad porque claro tengo el hijo allí.
   %tra: to me this is an eternity because you see I have the son there.
   04 *RES: <claro> [!] <qué edad tiene> [¿].
   %tra: <of course> [!] <how old is he> [¿].
   05 *MER: ahora tiene trece.
   %tra: now he is thirteen.
   06 *RES: <trece> [¿] <ui qué tal está> [¿].
   %tra: <thirteen> [¿] <ui how is he doing> [¿].
→ 07 *MER: está muy bien ahora estaba hablando con él está muy bien.
   %tra: he is pretty well I was now talking to him he's doing pretty well.
   08 *RES: sí¿
   %tra: yeah¿
→ 09 *MER: pero claro tu sabes me echa mucho de menos y +...
   %tra: but of course you know he misses me a lot and +...
   10 *RES: +^ claro.
   %tra: +^ sure.
→ 11 *MER: y yo también.
   %tra: and me too.
   12 *RES: claro.
   %tra: sure.
→ 13 *MER: bueno pero al menos con el internet que por ahí es una vía rápida.
   %tra: well but at least with the internet that is a quick way out.
   14 *RES: claro # <tenéis que quedar a una hora o no> [¿] -, aquello que le dices +"/.
   %tra: sure # <do you have to arrange a time or not> [¿] -, that's when you tell him +"/.
   15 *RES: +" mañana a las tres sí.
   %tra: +" tomorrow at three yes.
→ 16 *MER: sí por ejemplo con el internet sí porque él allí en casa no tiene internet # bueno tiene pero le quité ya la línea de internet porque ya se metió mucho en eso y entonces ya se estaba descuidando los estudios.
   %tra: yes for example with the internet we do because he doesn't have internet there at home # well he does have it but I already cut off his line because he was getting too much into it already and so he then was already neglecting his studies.
   17 *RES: vaya es que además en esa edad es más difícil,, <no> [¿] <controlarlos> [¿].

%tra: I see and besides at that age it is more difficult,, <isn't it> [¿] <to supervise them> [¿].
→ 18 *MER: <demasiado> [!] una llamada y ya yo sé que tengo que esperar.
%tra: <you tell me> [!] a call and then I already know I have to wait.
19 *RES: ah vale.
%tra: ah ok.

Merche explains that she has not seen her son for four years (line 2): an eternity (line 3). She openly voices her feelings of nostalgia and distress at the distance (line 11) and also her son's (line 9). Being a transnational mother is by no means an easy enterprise. However, against the idea of migrant women 'losing control' and becoming placeless inbetweeners, she has indeed managed to be 'here and there' (Diminescu, 2008: 572; Hondagneu-Sotelo & Avila, 1997: 548; Vertovec, 2001: 575). In short, she, like James, practises *simultaneity* without contiguity (Castells, 2004; Lewin Tapia, 2004; Peñaranda-Cólera, 2011). She has just been 'talking' to her son in an online conversation (line 7), she has just 'put him offline' because he has been neglecting his studies (line 16) and, as a mother, she found it harder when kids become adolescents (line 18) – note that despite the fact that she uses the distal deictic locative 'there' (at home; line 16) she keeps locating the deictic centre of her actions in the 'here and now' of the neighbourhood. This suggests that they have been able to jointly normalise their transnational mother–son ties via ICT. Merche relies on the internet (line 13), which, in her case, eases the scheduling and the price of daily connection, despite the five-hour difference in time zones. Also, they employ the 'beeping system', which, as we have already seen, consists of making a missed intentional call as a sign that they are both ready to go online without having to spend extra money on computer services or on calls to check the interlocutor's availability (line 18). Merche's ICT-mediated transnational practices also show that the discourses on transnational mothers' need for 're-socialisation' are unfounded and unrealistic. For many migrant women like Merche, transnational motherhood seems to be experienced as part and parcel of the workings of the transnational family configurations of the global era, through which they navigate and, in general terms, clearly habituate.

To conclude, transnational migrants today have gained an unprecedented degree of individual autonomy in managing their budgets, in exerting selective sociality and in adjusting the volume of their wanted and unwanted calls, despite some difficult realities, like the experience of enduring futile, unfinished conversations with relatives and friends. Therefore, in and through the *locutorios*, they have uniquely developed tactics that allow them to share and redistribute networking capital among their families, successfully collectivising such a precious resource for transnational life.

Migrant ICT users can now do so from the margins of the formal circuits of commerce, partly thanks to the establishment of migrant-tailored initiatives like the *locutorio* shops, which have become crucial for the articulation of the communicative practices that I have analysed in this chapter. To start with, *locutorios* have made particular technological gadgets readily available for all sorts of migrant populations who would otherwise have much more restricted access to ICT-mediated transnational communication, with successful legal formulas that allow for the connection of undocumented and disconnected people. Also, these call shops offer the whole range of telephone and computer services all at once, and they sell them in a way that suits the clients' real need to use them strategically, in complex combined manners, one after another, or even simultaneously.

Besides, *locutorio* entrepreneurs (perhaps out of commercial interest) are particularly sensitive to the socioeconomic restrictions that marginalised populations endure. Frequently used mass-consumption products and services are generally inexpensive (especially international calls) and allow for clients' close monitoring of very limited budgets. (Cabins, computers and money transfers all have systems that tell users exactly what they are spending.) And, of course, *locutorios* provide an excellent informal technology site at which to learn from the communicative practices of other users, to exchange and develop new 'tricks' (like the money-saving 'double SIM' tactic) and, in short, to share, maintain and expand one's transnational social networking capital, at the margins of mainstream society.

In this regard, *locutorios* have made an important contribution to propelling the technology-geared profound *mobile-isation* to which the quote that opened this chapter alludes: a profound ICT-mediated social revolution whereby, through a series of mundane resistance practices subverting given market rationalities, migrant networks can collectively appropriate, transform and cater for information and communication technology, via the linguistic practices and literacy capitals that are unfolded in the following chapter.

5 *Locutorio* Voices: Language and Literacy in Migrant-Regulated Discursive Spaces

> *In Barcelona* locutorios *[...] are almost always run by Pakistanis who do not know much Spanish. It seems that they are only trained to say 'an hour, a euro' and then they don't tell you 'please go to computer number six'; instead they only tell you 'six'. 'Six what¿' I asked him the first day, and the all-too-typical Pakistani working there at the moment looked at me in distress, his face saying 'I don't get a word from this guy'; his neurons were uselessly trying hard to understand my message without success. 'Chill out Pakistani', I told him, 'I'm moving to computer number six because I guess that this is what you want me to do, but then don't look at me with this face as if you had been scolded like a boy; and don't make such a fuss about this, in here we are not fighting'. Next time I told him I only wanted internet connection for half an hour, and asked him how much it cost. The Pakistani put the same scolded face again. 'Look Pakistani', I told him, 'an hour costs a euro [...], half an hour costs 50 cents, f-i-f-t-y c-e-n-t-s Pakistani, ok¿'51*
> José Joaquín López, El locutorio del pakistaní
> (blog at http://www.anecdotario.net, 11 April 2006)

This post, written by a Guatemalan migrant writer on his own public blog site, illustrates the complex social identities and the alternative linguistic hierarchies that may be observed in discursive spaces like the *locutorios* around the Barcelona area, where a radical heterogeneity of migrant populations with a wide range of different communicative frames, language repertoires and literacy capitals meet – and compete.

By analysing similar intercultural encounters that occurred inside the *locutorio* of El Paso, in this chapter I investigate the migrants' bottom-up construction, organisation and practice of multilingualism in a self-regulated space where the habitualised language norms of the host societies can be suspended, challenged and subverted. I first examine the functions and roles that both allochthonous and local languages play inside *locutorios*, and then explore the diverse social meanings and values attributed to them there. Likewise, I enquire into which migrant groups get to define what linguistic behaviours are legitimate or non-legitimate (i.e. sanctioned and silenced, in Bourdieu's sense; Bourdieu, 1991); why and under what circumstances, for a myriad of different allied or rival transnational populations. In other words, I focus on the inter- and intra-group language-triggered power dynamics that regulate the social structuration of various migrant networks in Catalonia today.

In the first section, I explain that *locutorios* mobilise a 'transnational appeal' which consists of a welcoming multilingual front-stage, but then show that, in reality, newer sociolinguistic regimes emerge which ultimately do not allow for all languages to take the floor in public. Instead, a (trans)linguistic unicity in Spanish as the public language of 'ethnic' businesses in Catalan metropolises is explicitly demanded by both habitual and non-habitual Spanish-speaking migrant workers and clients, at least in urban neighbourhoods where the linguistic market is secured in Spanish.

I argue that this is so because the Spanish language today is the material and symbolic capital with which most transnational migrant populations organise their relocation in, and insertion into, the Barcelona area (Alarcón & Garzón, 2011a; Codó, 2008; Codó & Garrido, 2010; Moyer, 2011). Thus, Spanish has become a crucial asset for migrants seeking incorporation into the (informal) labour market, accessing welfare services and making claims of 'proper citizenship' and of 'integration' into the host societies. It has also come to be the dominant lingua franca across migrant groups (Alarcón & Garzón, 2011c: 128; Corona *et al.*, 2013: 183; Hernández-Car, 2011: 116; Newman *et al.*, 2013: 200; Pavez Soto, 2011: 77; Pujolar, 2010: 231).

Under these circumstances, I show that languages other than Spanish tend to be relegated at the back-stage, or else pass unnoticed and become silenced in public, as occurs with Urdu, Panjabi, Pulaar or Darija as well as with Catalan, which, still laden with some ethnic and middle-class connotations, seems not to have entered the *locutorio* spaces. Of course, there is a competition, too, over what particular type of Spanish counts as the legitimate code of these institutions. This results in a frequent language battle that, as illustrated above with the anecdote in the blog post, most prominently arises between, on the one hand, diverse Latin American Spanish-speaking *locutorio* users who, by virtue of their linguistic background, position themselves as the authentic 'native speakers' of the Spanish language and the *locutorio* workers of Pakistani origin, on the other, whose 'non-pure' Spanish with multilingual transcodic marks gets delegitimised, mocked or even derided as 'non-talk'.

And yet, despite these explicit attempts to discredit multilingual 'interference' on the part of some migrant clients, I argue that what has actually colonised the *locutorios'* public floor is a type of fluid, translinguistic

Spanish rooted in linguistic hybridity, heteroglossy and technology-related alternative vernacular literacies. This is so regardless of the fact that, simultaneously, and as a consequence of their systematic decapitalisation, many non-schooled, non-Spanish-dominant migrant populations genuinely try to enculturate themselves into 'correct' Standard Peninsular and Latin American Spanishes, and so deploy a series of self-correcting and self-disciplining practices both in their written notes and in their actual oral performance.

In the second part of this chapter, I examine this unique, flexibilised transcodic Spanish with orality traits and unconventional spelling. I first show that it is an amalgamation or 'cluster' of Peninsular and Latin American Spanishes as well as of allochthonous codes; it is a fully fledged transcodic 'zero point' or ecumenical lingua franca which eases intercultural communication (Blommaert, 2012; Jacquemet, 2005, 2010). As such, I argue that it indexes membership and belonging to a truly migrant-tailored institution, because it is used by most *locutorio* users to protect resources within their self-regulated space, out of reach for non-migrant populations, particularly in the written mode. Also, I suggest that this non-standard multimodal code is collectively mobilised by migrants to publicly vindicate a legitimate transnational 'voice' (Jaffe & Walton, 2000: 562) in Spanish, one with which to claim the right to the resources of the Catalan urban floor.

At the end of the chapter, I take a step further and reflect upon what the transnational migrants' contradictions and ambivalences in their ideologies and in their practices may tell us about their linguistic incorporation into their host societies. I suggest that these incongruities are both the product and the process of the *migrantisation* (that is, of the migrants' successful appropriation) of the Catalan local economy, under the conditions of late capitalism. I conclude that this translates into the successful emergence of counter-hegemonic, anti-standardising vernacular language practices which speak of a powerful confluence of many different worldviews and of convoluted transnational mobility trajectories, today *inscribed* in the multi-lingualisms that have manifestly taken roots in the linguistic landscapes of Catalan neighbourhoods, in an unprecedented manner. On the other hand, though, I show that this *migrantisation* also speaks of the migrants' counterproductive remobilisation of old sociolinguistic hierarchies (like the establishment of a Spanish monolingual floor, or the 'backgrounding' of migrant allochthonous languages) and of heightened 'internal' fights for access to vital resources for transnational survival. This suggests that these populations paradoxically end up reproducing and exacerbating situations of social difference, inequality and exclusion mediated and manifested in and through language, from *within* their own transnational networks.

The Organisation of Silenced Multilingualisms in a Spanish-Unified Floor

Locutorios are given appealing names such as 'Locutori Global', 'Intercontinental', 'Intercomunicaciones' and 'Locutorio Mundial' (Páginas Amarillas, 2012), which, at the front-stage, evoke mobility and transnationalism. In Catalan towns I saw other signs that much more explicitly emphasise the multilingual character of these places, like 'Phone's Mobil y Internet', 'Locutorio Hello Hola' and 'Amigophone'. Thus, overcrowded with pieces of information in many different languages, *locutorios* are highly multilingual discursive spaces, at first sight.

This is partly so because some of their commercial staging is imported from the largely ineffective, tokenistic written *commercial multilingualism* of the Spanish telecommunications sector, which sells some of its products directly to *locutorio* owners. The presence of this commercial multilingualism is illustrated in Figure 5.1, which shows the seemingly 'multilingual' prepaid telephone card Habibi. Despite the fact that its name is written in Romanised Standard Arabic (it means 'my beloved' [male]), accompanied by an indication in English reading 'international call card', the instructions on the back (including the secret code, the customer services, the redial functions, the validity after first use, and so on) are actually all written in Spanish, except for one single access code which reads 'English payphone'.

Many of the real linguistic practices observed on the ground in these discursive spaces also show multilingual features, but these are of a very different kind, for they tend to be highly practical. In the oral mode, for instance, I recorded the systematised use of expressions like '*the recarga*' or '*el tecnición*', which show Spanish and English contact, and which have become the routinised, unmarked terms for 'the top-up' and 'the technician' by non-Spanish-dominant migrants of different linguistic backgrounds, both at the front-stage (for example, in interactions between *locutorio* workers and clients) and at the back-stage (among Pakistani *locutorio* workers).

These types of functional non-elite multilingual practices were also collected in the written mode, as seen in Figure 5.2, taken from the notebook that Shabbir, the construction worker from Pakistan, used for the autonomous learning of the Spanish language in the written mode. He (inaccurately – 'Wednesday' is actually written *miércoles* in Standard Spanish) copied a single word ('Wednesday') in Spanish and then wrote its meaning in Urdu in the Urdu script[52] and included the English translation below, again in the Roman alphabet, which, he claimed, helped him memorise the targeted Spanish word in the written mode.

Locutorio Voices 111

Figure 5.1 Tokenistic commercial multilingualism in *locutorios*. Pictures of the front and the back of a Habibi call card, taken by author (with selected details removed). *Locutorio* in Vallès Occidental, 15 October 2008

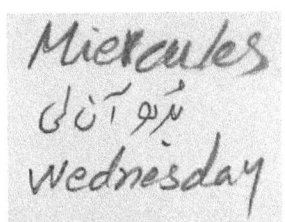

Figure 5.2 Non-elite written multilingual practices in *locutorios*. Shabbir's notebook. Picture taken by author. *Locutorio* in Vallès Occidental, 28 July 2008

Despite these situated multilingual practices, though, the idealised view of *locutorios* as almost ontologically multilingual institutions should be problematised, for, in fact, they are actually regulated in and through a local normativity regime which ultimately fosters the habitualised use of Spanish as the public language of the place. Informants Abou, from Guinea, and Abde, from Morocco, provide an example of the ways in which *locutorios* are constructed as Spanish monolingual zones. In Excerpt 5.1 they reflect on the linguistic behaviours of transnational migrants and on the hierarchies of languages that they both claim to have found there.

Excerpt 5.1. Spanish uniformity and monolingual ideologies at the locutorio

@Location: 1 September 2008. Bar near a *locutorio* in El Paso. Vallès Occidental.
@Bck: Abou (ABO) and Abde (ABD) present their views on the sociolinguistic orders of the *locutorio* to the researcher (RES). Abou emphasises the overarching use of Spanish inside the place, and while Abde does acknowledge the linguistic diversity, he then seemingly reproduces a monolingual line of thought.

```
   01 *RES:   <la gente> [¿] cuando vas al locutorio <la gente qué te
              habla> [¿].
      %tra:   <people> [¿] when you go to the locutorio <what do
              people speak to you> [¿].
→  02 *ABO:   me habla castellano.
      %tra:   they speak to me in Castilian.
→  03 *ABD:   +^ cada uno en su idioma.
      %tra:   +^ everyone in his own language.
   04 *RES:   <cada uno en su idioma> [¿] pero cuando vas a pedir o
              preguntas # o aquí al bar o +...
      %tra:   <everyone in his own language> [¿] but when you order or
              ask # or here in the bar or +...
      %add:   To Abde.
→  05 *ABD:   +^ castellano.
      %tra:   +^ Castilian.
→  06 *ABO:   castellano.
      %tra:   Castilian.
   07 *RES:   castellano¿
      %tra:   Castilian¿
→  08 *ABD:   sí.
      %tra:   yes.
```

→ 09 *RES: <y cómo es que:> [/] <y cómo es que es eso> [¿] # y: <por qué no el catalán> [¿].
%tra: <and how come:> [/] <and why is this so> [¿] # a:nd <why not Catalan> [¿].
→ 10 *ABD: hombre que en España el castellano -, el idioma el castellano.
%tra: look in Spain Castilian -, the language the Castilian.

When I ask Abou what languages people speak to him in the *locutorio*, he answers that they speak Spanish (in line 2). Abde, who is always ready to pick up on Abou's ideas, in a quick uptake provides a different answer by claiming that, there, everyone speaks 'in his own language' (line 3). When it comes to the languages used for other daily routines, they both agree that the public language of their interactions – the frame – is Spanish, too (lines 5, 6 and 8). When I ask them why this is so (in fact, I had arranged the interview to find out about the role of Catalan in El Paso; hence my question in line 9), Abde claims that this is so because, as a matter of fact, the language of Spain is Spanish (line 10). Thus, the *locutorio* is presented by its users as a Spanish-using space which 'facilitates' intercultural communication among linguistically diverse migrant groups. Following this line of thought, those individuals who are identified as speakers of different allochthonous codes shall use languages other than Spanish only at the back-stage, for more private, intra-group matters. In this way, both informants participate in, and actually reproduce, the apparently logical but in fact highly ideologised and politicised idea that the official language of the nation-state, Spanish, should be the natural, taken-for-granted language of the public arena. The autonomous region where these informants settled is talked about as a part of a single unified territory, Spain, not as Catalonia. The co-official language of Catalonia, Catalan, is not even mentioned by the informants; it is brought up by the researcher. At the same time, these highly multilingual migrants (I later learnt that Abou speaks Pulaar, Susu, Manya, French, English, Spanish and some Catalan, and that Abde commands Darija, Spanish and some spoken French, too) reproduce the *monolingual norm* (Heller, 2001). That is, they assign or associate each *locutorio* user with one single language according to his or her linguistic origin (normally the nation-state language of the speaker's place of birth), even when they know that most users are multilingual speakers like them (this monolingual-norm ideology is seen in Abde's comment that 'everyone speaks in his own language'). This prevailing regime of thought translates into the emergence of transnational institutions of migration whose linguistic diversity is ultimately constructed around a discursive space regimented in (translinguistic) Spanish, as I shall now illustrate with a detailed analysis of the values and functionalities of Spanish, English, Catalan and other allochthonous minority languages, presented below in this order.

Spanish

Naeem, the *locutorio* worker, could have chosen to set up the language for conducting money transfers in 'English UK, English US, French France, German Germany, Italian Italy, Spanish Mexico and Spanish Spain' (as advertised on the screen of the main computer, reproduced here verbatim). For him, English would have been the commonsense option, because he found it easier to read and write in the language in which he had been schooled in Pakistan. In fact, I recorded him saying that *'cuando yo me escribo siempre penso como English'* ('me when I write I always think like in English'). I also observed that when he had to make similar choices for private matters (for example, when he had to set up the language on his personal mobile phone) he still systematically chose English. And yet, at the workplace he chose 'Spanish Spain', or Standard Peninsular Spanish, instead.

When I asked him about his choice of Spanish, Naeem stated that he did not find it appropriate to use another language in what he understood to be a Spanish public business, where there already was a smooth, uniform floor for work: top-up or fax receipts, bank reports, invoices from the software and the telephone companies, and bills from the distributors of international call cards, for instance, were all provided only in Spanish. Besides, the language of technology and the semantic field of numbers (basically, prices, hours, postcodes and schedules) had already entered the *locutorio* in Spanish, too. Thus, routinised Spanish expressions such as *mañana* ('tomorrow'), *cambio* ('change'), *cabina* ('cabin') and so on were systematically adopted by habitual and non-habitual Spanish users, at the front-stage. Moreover, Naeem believed that his choice of Spanish was perfectly consistent with the post-modern international appeal that he wanted the *locutorio* to project, in a sense reinforcing the idea that the Spanish which circulates in *locutorios* is truly transnational. As he put it to me when broaching this topic: 'First Spanish. Spanish is the second world language'.[53]

In fact, the use of Spanish was fostered by not only Spanish-dominant but also non-Spanish-dominant *locutorio* workers and habitual clients alike, regardless of whether they were Spanish residents (many being of southern Spanish descent) or foreigners, as shown in Snapshot 5.1, which presents two communicative events illustrating the informants' linguistic ideologies around, and their investment in, a Spanish-only local normativity regime, taken from my fieldwork notes from 30 August and 16 September 2008.

Snapshot 5.1. The demand for a Spanish public floor both by Spanish-dominant and non-Spanish-dominant clients

At night, Jenny, Naeem's best friend, always stands by the *locutorio* door with a cigarette, and I every now and then follow suit. Naeem, dying for a smoke, also follows her whenever the *locutorio* is not crowded. Then busy Rachid, still being trained as a *locutorio* worker, approaches Naeem in Urdu/Panjabi in a hurry, and Jenny interrupts them with the warning 'Hablar en español, que estamos en España' ('Speak in Spanish, we are in Spain'). Born in El Paso, she explains that it is the Spanish language which will help them get into the real workings of the neighbourhood, especially at the workplace – following the 'integration through Spanish' ideology in which she, like them, genuinely believes.

A young Moroccan couple squeezed into one computer desk start an argument, in a code which I could not understand but which, I guess, must be Darija. Naeem does not seem particularly surprised about these sorts of fights, for, he says, they are habitual in a space like this. After a while, she suddenly stands up, grabs her handbag, and, visibly annoyed, shouts to her partner (in Spanish) '¡No me hables en árabe, coño, que estamos aquí!' ('Don't talk to me in Arabic, God damn it, 'cos we're here!') before leaving the place alone.

Under these circumstances, the use and command of Spanish has become a *public barometer* to check the level of 'adequate integration' among migrants themselves, not only in the *locutorio*, but also in the neighbourhood, and, more generally, in their host societies. Spanish is thus a form of symbolic capital whereby migrant individuals are granted, and grant, some degree of respect, legitimacy and even prestige to transnational migrant 'others' (Barth, 1969). Thus, this 'integration through learning Spanish' ideology, which, as Jenny shows in Snapshot 5.1, is shared with Spanish residents as well, is deep-seated across most migrant groups, who see a command of Spanish as highly necessary to access the services of *locutorio* businesses.

Apart from the few Spanish-dominant local residents of Spanish descent who visit the *locutorio* regularly, it is the diverse groups of migrants of Latin American origin who are attributed (by themselves as well as by others) the Spanish linguistic capitals considered the unmarked, taken-for-granted *norm* in *locutorios*. This is so because, contrary to what may happen in non-migrant-regulated discursive spaces, in this particular institution the Spanishes of these different groups are all usually considered to be legitimate codes. Therefore, both the Spanish residents' non-standard southern Spanishes (with traces of Catalan) and the migrants' Latin American Spanishes enjoy substantial social value, because they are all categorised as codes which are mobilised by 'native speakers'.

Nonetheless, there do tend to be isolated competitions over linguistic capitals between Spanish residents and migrants from Latin America concerning whose accent is the more legitimate (as attested, too, by Corona *et al.*, 2013: 183; and by Trenchs-Parera & Newman, 2009: 517, 521). I observed this, for example, when they, respectively categorised themselves as *'españoles'* and as *'latinos'* or *'sudamericanos'* when seeking (and competing for) a roommate (see Chapter 6). However, upon the *locutorio* desk, I recorded that *both* Peninsular-sounding *móviles* and Latin American *fonos* (for mobiles), *bolis* and *lapiceros* (for pens), *chicles* and *gomas de mascar* (for chewing gum) or *ordenadores* and *computadoras* (for computers) are accepted as the basic 'translinguistic Latino repertoire' (Corona *et al.*, 2013: 186) for transnational survival. Interestingly, this multi-voiced *locutorio* language is discursively constructed, indistinctly, as both *español* and *castellano* by both migrant and non-migrant populations, which may indicate that it is actually a set of transnational fluid Spanishes (Abou and James call it *castellano*, and Jenny and Naeem call it *español*, for instance).

Non-dominant Spanish-using *locutorio* workers and users, particularly of Pakistani and Moroccan origin, take these polyphonic routine expressions as the model forms from which to enhance their command of Spanish, while Spanish locals and, perhaps more insistently, Spanish-dominant speakers from Latin America take advantage of the social value attributed to their linguistic capitals to publicly position themselves as 'better' than them (for example when complaining about given *locutorio* norms established by Naeem). In turn, they attribute to themselves more expertise in matters concerning migratory trajectories. The privileged position of this flexible hybrid Spanish as a lingua franca and the particular tug of war between 'non-native' Pakistani and 'native' Latin American migrants is illustrated in Snapshot 5.2, which gathers some fieldwork observations from 30 June 2008.

In some cases, some of the Latin American *locutorio* users have become linguistic activists who take it as a personal commitment and as a collective moral obligation on the part of the Spanish-speaking 'nations' to support the use of the Spanish language not only as the public language of Catalonia but also as *the* global language of technology. This linguistic activism is

Snapshot 5.2. 'Native' translinguistic Spanishes as the legitimate public codes of locutorios

When clients from Latin America seek to subvert the *locutorio*'s order, they tend to attack Naeem by making fun of his oral expertise in Spanish, in turn positioning themselves as more legitimate than the non-Spanish-dominant clients. Today, for instance, a young man with a Latin American accent who had been asked to move to cabin number four got annoyed and publicly accused Naeem of not knowing how to pronounce the 'Spanish' 'r' sound correctly, dispossessing him, in passing, of manly capital. 'Cuatrrrro…. ¡Habla como un hombre, coño!' ('Fourrrr… Speak like a man, damn!'), the client complained while lazily switching on computer number four. This reminds me of the day when Naeem was talking to the Bolivian neighbour who similarly discredited some of his Panjabi-sounding Spanish expressions with an *'¡Habla más claro!'* ('Speak more clearly!'). No wonder why Naeem keeps repeating (in Spanish!) that he has to learn Spanish because he doesn't speak it well (he also frequently asks me to 'correct' his SMSs in written Spanish, too). '¿Lo ves? No les gusta!' ('See? They don't like it!'), he told me on that particular day.

illustrated in Excerpt 5.2, where James, the middle-aged protest poet from Cuba, defines and emotionally legitimises the worldwide dominion of the Spanish language as a vital 'global currency': as both a necessary economic asset under the present-day conditions of the globalised new economy and as a powerful international language useful for successful intercultural communication across different ethnolinguistic groups (note that the differences between peninsular Spanishes and other Spanishes from Latin America are left unmentioned, and that the language is, in this case, called *castellano*).

Excerpt 5.2. Linguistic activism at the locutorio: *The defence of the worldwide dominion of global Spanish*

@Location: 3 August 2008. *Locutorio* in El Paso. Vallès Occidental.
@Bck: In a conversation about the role of languages in the telecommunications world with the researcher (RES), James (JAM) defends the idea that Spanish is and should be a dominant language of communication in Catalonia, and worldwide as well.

01 *JAM: dentro del mundo de las telecomunicaciones ya que tú lo estás abordando.
    %tra: within the telecommunications world now that you raise it.
02 *RES: +^ claro.
    %tra: +^ sure.
→ 03 *JAM: pues un lenguaje común del uso de: muchos millones de personas en el mundo que es una riqueza que tiene España como es el castellano -, pues resulta relevante: # es una divisa que tiene esta nación.
    %tra: well a common language of use o:f many millions of people in the world which is a wealth of Spain as much as Castilian is -, well it turns out to be importa:nt # it is a currency that this nation possesses.
04 *RES: es una divisa؟
    %tra: is it a currency؟
→ 05 *JAM: el castellano sí <entiendes> [؟] que cada vez son más las personas que: # mira ya más del veinticinco por ciento <de> [/] de Norte América que ahí xxx la nación del mundo.
    %tra: Castilian yes <you get me> [؟] that there are more and more people all the time who: # look already twenty five per cent <of> [/] of North America that there xxx the nation of the world.
06 *RES: aha.
→ 07 *JAM: allí casi todo el mundo habla castellano <y> [/] y se estima que en diez o quince años <habrá> [//] la mitad de la población americana hablará castellano.
    %tra: there almost everybody speaks Castilian <and> [/] and it is estimated that in ten or fifteen years <there will be> [//] half of the American population will speak Castilian.
08 *RES: aha.
→ 09 *JAM: y se estima que en en medio siglo o un siglo ya va a ser el idioma más hablado del mundo.
    %tra: and it is estimated that in in half a century or in a century it'll be the most widely spoken language in the world already.
→ 10 *RES: y qué vamos a hacer con los móviles؟
    %tra: and what are we going to do with the mobiles؟
→ 11 *JAM: esto no va a relegarse a: porque nunca (؟) pero como los móviles hablan castellano pues no nos queda más remedio,, <no> [؟][=! laugh].
    %tra: this won't be relegated to: because never (؟) but since mobiles speak Castilian then we have no choice,, <right> [؟][=! laugh].

```
    12  *RES:   ah.
→   13  *JAM:   esta es la fuerza impresionante de <de> [//] del castellano
                <tú lo sabes> [!] <no te estoy contando ninguna historia>
                [!].
        %tra:   this is the powerful strength of <of> [//] of Castilian <you
                know it> [!] <I'm not making up a story here> [!].
    14  *RES:   no no.
        %com:   Unconvincing tone.
→   15  *JAM:   está ahí a la orden del día de internet que están las
                comunicaciones -, <infórmate> [!] mira los informes de
                Naciones Unidas eh <de lo> [//] de las distintas lenguas
                que se hablan en internet # el crecimiento de ellas <el> [/]
                el crecimiento del castellano <pero con una fuerza> [!].
        %tra:   it's there daily on the internet where communications are
                -, <check it out yourself> [!] look at the United Nations
                reports eh <on the> [//] on the different languages that are
                spoken on the internet # on their growth <the> [/] the
                growth of Castilian <but with such a strength> [!].
    16  *RES:   ya.
        %tra:   sure.
→   17  *JAM:   <no> [/] <no> [/] no es que quiero relegar el catalán <es
                que es una realidad que se impone en el mundo> [!].
        %tra:   <no> [/] <no> [/] no it's not that I want to relegate
                Catalan <it's that it is a reality which is being imposed
                around the world> [!].
    18  *RES:   sí.
        %tra:   yeah.
→   19  *JAM:   si renegaran de esta riqueza -, mira nosotros nos estamos
                comunicando perfectamente.
        %tra:   if they repudiated this wealth -, look we are now communi-
                cating perfectly.
    20  *RES:   claro.
        %tra:   sure.
→   21  *JAM:   <vamos a renegar de esta riqueza> [¿] <no> [!] # ahora
                tú tienes tu lengua materna -, está bien que tú la conserves
                que tú la quieras que no muera porque pues eso es una
                riqueza cultural y está bien que en tu entorno familiar en tu
                entorno común pues alguien que se eduque incluso dentro
                de Cataluña <en> [/] <en> [/] en los colegios -, pero nunca
                que se relegue el castellano.
        %tra:   <are we going to repudiate this richness> [¿] <no> [!] #
```

now you have your mother tongue -, it's ok that you keep it that you love it that it doesn't die out because well that is it's a cultural richness and it's ok that in your family environment in your common space then even somebody being educated within Catalonia <in> [/] <in> [/] in schools -, but never should Castilian be relegated.

%act: James pauses. He may be reading the researcher's silence as non-agreement.

→ 22 *JAM: sí es una lengua dominante porque sé la teoría y porque ya he vivido más de tres años anteriormente aquí [en Cataluña] y yo he tenido muy buenas relaciones con catalanes # me han hablado de la lengua dominante que tiende a relegar a la nativa etcétera y son teorías pero yo pienso que si la gente estudia en el colegio nunca se va a morir y es la lengua que mamas de tu mamá de la teta de tu madre.

%tra: yes it's a dominant language because I know the theory and because I've lived here [in Catalonia] for more than three years previously in here and I've had very good relationships with Catalans # they've told me about the dominant language that tends to relegate the native one etcetera and these are theories but I think that if people study at school it'll never die out and it's the language that you are breastfed by your mum from your mum's breast.

[...]

→ 23 *JAM: ahora lo que es una verdad ineludible es esto del castellano imponiéndose en el mundo y que es una riqueza que hay que cuidar y <es un> [//] es una delicia española y de España porque mal que le pese a quien le pese <esto es España> [!].

%tra: now but what is an unavoidable truth is that Castilian is growing in strength around the world and that it's a wealth that should be taken care of and <it's a> [//] it's a Spanish delight and of Spain because whether they like it or not <this is Spain> [!].

James starts off by defending the use of Spanish in the global system of communication, on the grounds that in 50 years it will be the most widely spoken language in the world (line 9). He states that it is a currency of Spain, which he defines as a 'nation' (line 3), and he repeatedly uses big numbers and estimates (for example in lines 5 and 7) as a 'scientific' authoritative basis to justify these claims, at times (when I look unconvinced) using the *argumentum ad verecundiam* that powerful modern entities and supranational authorities like the United Nations say so (lines 13 and 15). This, to him, grants the homogenised Spanish-speaking 'imagined community' of users (Anderson, 2006 [1983]) justification and accountability for promoting the use of Spanish-only policies in the telecommunications world, and for leaving its expansion unproblematised. When I, challenged and uncomfortable, ask him what will happen to mobile phone communication (line 10), he states that we have no choice but to follow the economic market rationalities which bet on dominant linguae francae in this capitalist world (lines 11 and 17): 'business is business', he added in English later on.

Since his explanations might sound 'politically incorrect', he then takes a more civic road and, mobilising the classic neoliberal rhetoric around 'inclusiveness', presents the Spanish language as a 'symbol of democratic harmony' (Del Valle, 2006: 32) and as a 'richness' for global contacts (line 21), a wealth that we, as transnational post-modern citizens, cannot refuse, for it serves the sort of intercultural exchanges that we had in El Paso (line 19). In his globalist discourse, Spain's internal minority languages such as my 'mother tongue', Catalan, are appropriate within private spheres of my 'family environment' (that is, among those considered to be 'autochthonous Catalans'), but Spanish should come first (line 21). Discourses of minority language rights claiming that this linguistic comportment may relegate Catalan to a subordinate place do not apply (he says that this is simply the dominant-language-relegates-the-dominated 'theory'), for, he claims, no one loses the language with which one is breastfed and schooled – note the use of the emphatic adverb 'even' (line 21). In spite of his attempts to depict the Spanish language as a 'Spanish delight' (line 23) and to present himself as a respectful person with a legitimate knowledge of the Catalan cause (line 22), he abandons the neoliberal civic tone near the end, perhaps on realising that I was not convinced, and puts an end to our interview by using the same idea that Abde had previously expressed in a less insistent tone (in Excerpt 5.1, line 10): 'whether they like it or not, this is Spain!' (line 23).

English

Perhaps counter to common thought, the English language in this prevailing Spanish-only local normativity regime does not function as the taken-for-granted language of technology or social media gadgets, because it is not the language of ICT business among transnational migrant entrepreneurs, and because it has almost no economic value as an asset for transnational migrant employees seeking to incorporate themselves within the marketplace of *locutorio* businesses (this is so even for those who were schooled in it). Likewise, English has little instrumental usefulness as a lingua franca for transnational migrants' inter-group communication, and it is normally used only as 'transient capital' (Codó, 2012: 311), that is, as a resource for circulation (not for relocation) which is occasionally mobilised to 'get by' in everyday worker–client interactions when intercultural communication does not run smoothly. For example, a *locutorio* worker might employ it in an isolated manner after an unsuccessful top-up transaction undergone entirely in translinguistic Spanish, by repeating the clients' telephone number in English in order to double-check that he understood it correctly, because many clients have a basic knowledge of the digits in this language.

And yet, English does become a useful resource for particular identity enactment and identity construction processes linked to the interplay of some ethnicity, class, gender, religion and age domains. The complex ways in which this global language is locally embedded, moulded, taken up and mobilised for identity self-ascription or for other migrants' heterocategorisation purposes inside *locutorios* are manifold. This is so because most migrants who visit them have had many different lived experiences and have developed divergent and even oppositional *structures of feelings* (Giddens, 1991) towards this particular language. These feelings range from a proud sentiment of belonging to a given social class, age group or religion to a post-modern rage against a colonial past, or to a sense of embodied ethnic exclusion, uprootedness and social disjuncture in the host societies.

For some migrants of Pakistani origin, for instance, English is a highly valued in-group sign of class and educational distinction and a marker of affluent family background. For informant Shabbir, whose wife in Kashmir is an English teacher and whose children are being educated in an elite English-medium private school, English indexes the upper-middle-class status of the wealthy urban intelligentsia, which includes business people, lobbyists, military officers and high-rank bureaucrats, in Pakistan and in Pakistani Kashmir (Mansoon, 2004; Paolillo, 1996; Rahman, 1997a, 2002). Shabbir strategically switches from Spanish to English as an act of identity, in order to 'do' class and to distinguish himself from other Pakistani migrants who, like Naeem, may well know English but do not belong to any Pakistani circuits of power and have not experienced any sort of educational or socioeconomic improvement. This was observed, for instance, when he once switched from Spanish to English in front of me, the 'English teacher', to contradict some of Naeem's ideas about the workings of the administration in Pakistan, thereby attributing to himself more knowledge of the place. Shabbir's reappropriation and occasional strategic use of English in public also makes him different from Sheema, the non-schooled and non-literate rival religious leader of the local Muslim community who was born in a Pakistani rural town with faulty electrical infrastructure. In this case, English acquires elitist connotations and seems to reinforce pre-existing allochthonous caste tensions and power dynamics, now imported and relocalised across transnational migrant groups in their own self-governed institutions, in Catalonia.

Simultaneously, English can become part of some migrants' enactment and construction of particular religious identities. For example, for Sheema, who whenever one of his compatriots switches to English states (in English!) 'I hate English!', this language is part and parcel of a neo-colonial order imposed by what he calls 'the new empire', the United States. The use of English in public, it seems, goes against his transnational project to contribute economically to his currently troubled Islamic Pakistan, a project which he organised together with other Pakistani Muslim persons networking, on- and offline, in Vallès Occidental, Barcelonès, the United Kingdom and Pakistan. Sheema rejects English and anti-Muslim discourses alike, just as he rejects non-Muslim political and military intervention in the Muslim world. This rejection of English is not so much about language or about foreign intervention *per se*, but about the religious connotations that, for Muslims like him, the English language now bears as the mobiliser of anti-Islamic discourses circulated worldwide after Al-Qaeda's attacks on the United States in 2001. This explains his active mission to 'erase' its use in the migrant-regulated spaces of the neighbourhood, particularly in what he considers to be a 'Pakistani' *locutorio* in El Paso (he reprimands Naeem whenever he plays English music there, for instance) and, of course, inside his own nearby bar (which he runs under someone else's name).

English can also be mobilised as a linguistic weapon to 'undo' given hetero-attributed (or other-ascribed) social identities, as well as to combat lived experiences of social inequality exerted by competing transnational migrants in public. For example, one day Naeem was spraying strawberry air freshener all around the counter, a daily practice during the summer in order to hide what he perceived as summer malodour, when a black man in sweaty work clothes who was queuing to pay (and who had previously used Spanish to order his cabin receipt, as usual) took it as a personal offence and reacted by changing code and shouting, in an African-sounding English: 'You're crazy man, you're always absent-minded!' He checked whether his T-shirt had been stained by the air freshener and then left quickly, visibly annoyed, while Naeem tried to explain to the rest of the clients (in Spanish) that the spray had not been directed at that particular client. I read this customer's code-switching to English as a way to complain about, and to reject, what he perceived as his other-ascribed, classed and ethnicised social categorisation as the 'smelly pauperised black worker' of the *locutorio*.

For other transnational *locutorio* users English may play a key role in the articulation of global youth identities. For some young male migrants of Pakistani origin, for instance, this language may become a marker of a transgressive, masculinised identity, such as that of belonging to the global 'hip-hop nation'. Rachid, the *locutorio* trainee from Pakistan, for instance, appropriated some hip-hop English slang when trying to readjust to his new life as an undocumented migrant. He always wore a wide sweatshirt and baggy trousers, and the first thing he did when he gained Naeem's trust at the *locutorio* was to substitute the river scenery on the computer screensaver for a picture of him DJ-ing with his break-dancing male friends. He is a hip-hopper who loves English lyrics that speak of resistance. When clients laughed at his still not very fluent Spanish (he had then been there for less than two weeks), he would pointedly ignore them by putting his headphones on and by starting to rap along with a recording, immersing himself in the English lyrics as a way to overtly contest this hetero-attributed Spanish non-competence. Rachid's mobile phone setting and ringtone were in English, too, and he met many transnational friends during work hours on rap blogs and hip-hop websites, virtual communities of music sharers who tend to use alternative, non-standard Englishes as their language of (virtual) interaction (Androutsopoulos, 2009; Codó, 2012; Pennycook, 2007).

Finally, English can also become a language which indexes a looser postmodern cosmopolitan identity. For Naeem, for instance, the English-singing Céline Dion is perfectly compatible with Spanish-singing Enrique Iglesias and Urdu-singing Atif Aslam. In fact, he claims his right to have songs by all three played on the main computer, and finds it very hard to satisfy everybody's ethnolinguistically driven musical tastes inside the *locutorio*. In his case, English indexes, as it does in Catalonia in general, an emotionally 'neutral' (Huguet, 2007: 32; Llurda, 2009: 125), 'external' elite language of worldwide connection devoid of strong sociopolitical connotations, possibly following the unproblematised assumption that, as Naeem put it to me (in English), 'English is number one in the world'.

Catalan

Contrary to the popular idea that transnational migrant populations have no knowledge of, or perhaps no interest in, Catalan, what I observed is that this language is generally understood, and that the majority of informants had developed some knowledge of it, particularly younger migrants, some of whom invested in Catalan–Spanish bilingualism. And yet, migrants tend to see it as a non-public language with scarce immediate practical use and with no instrumental impact on their daily lives. Thus, ideologically, the Catalan language, unlike Spanish, is conceived of as if it did not 'belong' to their linguistic repertoires and as if it did not fit into migrant-regulated transnational spaces like *locutorios*.

One of the first reactions that many *locutorio* users of non-Spanish-speaking background had when I enquired about the Catalan language was that they associated it with the language of the Catalan administration, particularly with the language of the paperwork required by the town hall, for which they could always find someone within their own networks who could help (in El Paso, it is a public school teacher who volunteers to fill in forms written in Catalan). For example, whenever I asked the Pakistani *locutorio* workers about this language, they always pointed to the *locutorio* licence, compulsorily hung on the wall, which was issued by the local municipality to certify that everything was in order (this was, in fact, one of the few public documents entirely written in Catalan that I could find there). This may indicate that, for many migrants, Catalan is an outside code symbolically used as proof of legality, temporarily adopted only to be accountable to the Catalan authorities. This construction of Catalan as an 'institutional language' is often linked to a stereotyped image of the 'social correctness', 'distinction' and 'high culture' of an imagined wealthier Catalan host society, which may well cause the migrants' rejection of its incorporation into their repertoires (Alarcón & Garzón, 2011a: 144; Bastardas i Boada, 2012: 82; Hernández-Carr, 2011: 118).

I found yet another non-realistic association: that of linking this language with a small rural group of authentic and ethnically 'pure' (i.e. monolingual) distant Catalan 'native speakers' unaffected by incoming migration (Trenchs-Parera & Newman, 2009: 529). The ways in which I was assigned a Catalan identity when speaking Catalan to my non-foreign Spanish-using neighbours born in El Paso, presented in Snapshot 5.3 (from my fieldwork notes taken on 30 September 2008), provide an example of these perceptions, which, in the *locutorio*, were shared with transnational migrants. (I was born within a Catalan-speaking family and was educated in a small Catalan rural school in an overwhelmingly Catalan-speaking town with no more than 1,500 inhabitants, in central Catalonia.)

Snapshot 5.3. The middle-class connotations of Catalan inside the locutorio

After having spent some time together, Jenny told me (in Spanish) that Catalan (my sort of Catalan, she clarified) was for 'los del puño agarrao' – the tight-fisted, in a teasing tone. She sees it as an ethnicised and classed language that belongs to those 'Catalan' 'others' stereotyped as haughty elite workers who do not mingle with the Spanish-speaking working-class masses to whom, she feels, she belongs. For Carlos, an 18-year-old who every now and then nicks some cigarettes from the *locutorio* worker Naeem and kills time and loneliness with Jenny, Catalan can be a language for mockery, at times tainted with humorous 'rurality traits'. When I introduced myself, he also laughed at my Catalan (which may probably have fewer traces of Spanish contact than theirs), repeating my name in a singsong voice, and striking his bottom with his unoccupied hand, as if riding a horse: '¡Maria, Maria!' He thought that I either came from the rural areas of the Pyrenees, or else from Vic (in central Catalonia), where, he says, 'they do not use Castilian and have a very broad accent' (that is, where people are imagined to speak some sort of 'authentic' Catalan devoid of Spanish interference). Rafa, also a heavy smoker hooked on Facebook, born and schooled bilingually in El Paso, put an end to that first encounter with a resolute 'No pasa nada, yo soy gitano catalán' ('No big deal, I'm a Catalan "gypsy"'), in Spanish, making it clear that I was not following the expected sociolinguistic local norms, but that he was somehow fine with it. After several evenings on the benches outside the *locutorio*, both Carlos and Rafa, like Jenny, found it unremarkable that I then spoke mostly in Catalan, and that they spoke mostly in Spanish, unselfconsciously, with a fair amount of messiness, and without mockery. What they did find unusual was that a Catalan-dominant person shared the spaces of El Paso in Catalan, for that, according to them, was indeed a marked exception. My reading of this is that the people of this working-class neighbourhood, where the linguistic market is secured in Spanish, feel that the other co-official language of Catalonia, Catalan, is an 'outside' external language, one which is not for them. These are the messages that enter migrant *locutorios* such as the one I investigated in El Paso.

These constructions of Catalan definitely have to do with the specific location that the informants of this study inhabited, since it has been shown that for non-Catalan speakers to incorporate, and to make public social use of, Catalan there has to be a critical mass of habitual Catalan users in the immediate discursive spaces that they occupy on a regular basis (Hernández-Carr, 2011: 107; Pujolar et al., 2010: 16). The *locutorios* that I observed are located in a predominantly Spanish-speaking area neighbouring Barcelona where the public use of Catalan is scarce. The *locutorio* that was the principal place of research is located in a zone in Vallès Occidental, which in the late 1990s, 'appeared largely untouched by linguistic [Catalan] normalisation' (Woolard, 2003: 92),[54] and which today bears witness to the partial failure of the Catalan government's language policies, which, since 1981, have tried to foster the social use of Catalan as the de-ethnicised, unifying habitual language of Catalonia, first among Spanish-speaking families of Spanish descent, and later among transnational migrant populations, too (Aymà Aubeyzon, 2012; Plataforma per la Llengua et al., 2008).

In this particular Spanish-dominant sociolinguistic landscape, many transnational migrants inside migrant-governed non-mainstream worlds end up sharing the idea that Catalan is a language relegated to the private space of the home, and which 'belongs' to unknown groups of people who have little direct contact with working-class neighbourhoods. Consequently, for many migrants in El Paso, Catalan is perceived as a non-public 'vernacular' language. This is illustrated by Shabbir and by his flatmate Yousaf in Excerpt 5.3, where they seem to extrapolate or equate the role of Catalan in the Spanish state with the 'regional' status of the Panjabi language in Pakistan.

Excerpt 5.3. The construction of Catalan as a second-place 'home' language

@Location: 22 August 2008. Bar near a *locutorio* in El Paso. Vallès Occidental.
@Bck: Encouraged by the researcher (RES), Shabbir (SHA) and Yousaf (YOU) talk about their 'home' languages and seem to equate the 'regional' status of Panjabi in Pakistan to that of Catalan in Vallès Occidental.

```
   01  *RES:  y: vosotros en qué habláis en urdu en casa¿
       %tra:  a:nd you what do you speak do you speak Urdu at home¿
→  02  *SHA:  sí: Panjabi.
       %tra:  ye:s Panjabi.
   03  *RES:  ah Panjabi!
→  04  *YOU:  es como catalán.
       %tra:  it's like Catalan.
```

→ 05 *SHA: +^ cada uno país tiene uno: <cómo se llama> [¿] como
catalán o como **Panjabi** y siempre en casa y hablando así.
%tra: +^ each one country has one: <what is it called> [¿] like
Catalan or Panjabi and always at home and talking like this.

Shabbir had told me that he normally spoke Urdu at home, so I had wrongly assumed that his flatmate Yousaf would also use Urdu to talk to him. However, Yousaf was not born in the Urdu-dominant Kashmiri area (he was of Panjabi-dominant, rural and non-literate Pakistani background), and so in their flat he mostly speaks Panjabi (line 2). Yousaf affirms that Panjabi is a language that works 'like Catalan' (line 4), and Shabbir, departing from his explanations, clarifies that this means that, in sociolinguistic terms, Panjabi is a relegated language for the home (line 5), separately employed from official alternatives. In Pakistan, Panjabi is a relevant but devalorised non-official regional language with no role in the administration or in the public schooling system, and, instead, it is English, Urdu and, though to a much lesser extent, Sindhi which are the official or national languages, coexisting along with another 70 or so codes that have little recognition (Lewis et al., 2011; Mansoon, 2004; Rahman, 1997b). What I infer from Shabbir's and Yousaf's approach to Catalan is that they may perceive it as a code with a backgrounded status employed only for intra-family communication. In their description, the inferiority, minoritisation or relegation of Panjabi and of Catalan and the supremacy of their more widespread respective nation-state languages are naturally taken for granted and, in fact, left unproblematised. This provides evidence that, in this migrant-regulated space, the local sociolinguistic hierarchies of mainstream society in Catalan urban areas, where the linguistic market is secured as overwhelmingly Spanish, tend to be not only reproduced but also exacerbated among migrants themselves.

The view of Catalan as an inferior 'homey' language may just be one end of the negative/positive values and of the non-useful/useful roles attributed to Catalan. As outlined above, many transnational migrant populations acknowledge that an investment in Catalan and in Catalan–Spanish bilingualism is needed in order to function in the Vallès Occidental area. This belief is most notably shared by those migrants who at some point accessed the (informal) Catalan local marketplace, like Shabbir, for instance, whose superior was a Catalan-dominant employer working in the construction sector. When I more directly ask him about the role of Catalan in the neighbourhood, he rectifies an earlier statement (Excerpt 5.3) and says that this language is indeed effectual and socially utilitarian. According to him, though, the 'problem' is that those whom he hetero-categorises as 'Catalans' tend to systematically switch to Spanish when addressing migrant populations, and so he has no opportunity to practise speaking Catalan, as he explains in Excerpt 5.4.

Excerpt 5.4. The perceived linguistic convergence to Spanish by 'Catalans'
@Location: 10 August 2008. Bar near a *locutorio* in El Paso. Vallès Occidental.
@Bck: The researcher (RES) asks Shabbir (SHA) about the use of Catalan in the town of El Paso in Vallès Occidental, which prompts comments about the apparently systematised linguistic convergence or accommodation to Spanish by people he categorises as 'Catalans'.

<pre>
   01 *RES: se oye catalán aquí en www¿
      %tra: is Catalan heard here in www¿
      %com: www is the town where El Paso is located.
→ 02 *SHA: también pero cuando siempre cuando nosotros hablamos
              <de> [/] de castellano ellos [los catalanoparlantes] siempre
              hablan castellano.
      %tra: as well but when always when we speak <the> [/] the
              Castilian they [Catalan speakers] always speak Castilian.
      [...]
→ 03 *SHA: en cinco seis años no he visto ni una persona hablar catalán
              <esto de quién es falta> [¿].
      %tra: in five six years I haven't seen a single person speaking
              Catalan <whose fault is this> [¿].
</pre>

Shabbir claims that, indeed, Catalan is also heard in public, in his mostly Spanish-speaking town (line 2). However, he argues that when migrants like him address 'Catalans' in Spanish, they always accommodate and reply in Spanish, not in Catalan (line 2). As I understand it, Shabbir's comments speak of the fact that Catalan-using local residents have seemingly habitualised the switch to (Peninsular) Spanish when talking to Spanish-dominant users, both to the descendants of people from other regions of Spain and to transnational migrant populations (line 3).

Before the large-scale incoming migration movements from outside the Spanish state, this systematised linguistic convergence to Spanish by habitual Catalan speakers addressing non-Catalan-speaking populations had become an unmarked choice during the 1990s in the metropolitan areas of Barcelona, where many families from southern Spain settled (Boix i Fuster, 1993: 39). This embodied disposition on the part of the Catalan locals was largely unconscious, and was employed even when it was obvious that the Spanish-using interlocutors understood or spoke Catalan (Boix i Fuster & Vila i Moreno, 1998; Pujolar, 2009: 100; Woolard, 2003). The accommodation norm was a way to establish and to interactively maintain the now old ethnolinguistic boundaries between *Catalans* and *Castellans* in Catalonia (Woolard, 1989). On the one hand, the 'Catalan' locals employed Catalan only among themselves as a way to gatekeep access to the language of belonging and entitlement, which ensured economic advancement in Catalonia, and they switched to Spanish as a way to approximate, and to show respect and deference to, newly arrived Spaniards (Pujolar, 2007c, 2009). In turn, the 'Castellans' who at that time refused or did not invest in Catalan were distancing themselves from the middle-class connotations and from the symbolic attachment of Catalan to a Catalan (i.e. non-Spanish) identity marker, to the detriment of their linguistic repertoires, and to the detriment of the Catalan language, recovering from 40 years of Francoist prosecution.

Today, though, the children and grandchildren of the former Spanish-speaking populations from other parts of Spain have been schooled in Catalan and are mostly Catalan–Spanish bilinguals. Besides, these younger generations do not generally conceive of themselves as belonging to two parallel, separate communities of speakers, for they build and cultivate newer, more cosmopolitan, neoliberal projects of the self based on multiple 'both–and' fluid identities (Woolard & Frekko, 2013: 134). For these bilingual social players, therefore, the Catalan and Spanish languages are losing some degree of relevance as identity markers, all of which is leading to a de-politicisation of language choice in public discursive spaces in Catalonia (Frekko, 2013; Pujolar, 2011; Pujolar & González, 2013; Woolard, 2013).

However, linguistic convergence to Spanish is now being re-mobilised by these heterogeneous local residents of the host societies, regardless of their ethnolinguistic background (and regardless of their various late capitalist Catalan and/or Spanish identity ascriptions), who now tend to switch to Spanish automatically when addressing transnational migrant populations (Codó & Garrido, 2010; Corona *et al.*, 2013; Hernández-Carr, 2011; Pujolar, 2009, 2010; Pujolar & Gonzàlez, 2013; Sabaté i Dalmau, 2010). Thus, most of these hetero-categorised 'Catalans', who, according to informants like Shabbir, accommodate to Spanish are, in reality, fully fledged Catalan–Spanish bilinguals who are mostly habitual users of the Spanish language at home and who may not necessarily see themselves as only 'Catalans' (Pujolar, 2011: 28; Pujolar *et al.*, 2010: 32). This indicates that the perceived current linguistic convergence to Spanish on the part of 'Catalans' may have been overemphasised, as seen, for instance, when Shabbir states, in a teasing tone (with the use of the word 'fault'), that 'in five[,] six years' he has not seen anyone using Catalan with migrant populations (in Excerpt 5.4, line 3).

The newer uses of the accommodation norm towards transnational migrants on the part of Spanish residents fosters the emergence of a renewed sense of an autochthonous–native versus allochthonous–non-native ethnolinguistic boundary between those considered to be 'Catalans' (that is, Catalan–Spanish bilinguals with a command of Catalan, mostly local residents) and those considered to be 'non-Catalans' (that is, those who refuse any investment in the Catalan language as well as those non-Catalan-using transnational foreigners) (Woolard, 2009). By switching to Spanish in a routinised manner, the diverse Catalan- or Spanish-born local populations are blocking the migrants' chances to access the bilingual capital which is necessary to advance socioeconomically in Catalonia, and they are not giving them the grounds for learning the language which also symbolises membership of, affinity to and incorporation within Catalonia (this is part of Shabbir's complaint). Simultaneously, non-Catalan-using migrants who, like James (in Excerpt 5.2), believe that this language is not for them are denying the actual practical functionality of Catalan in all domains of social life and are letting go of a type of capital which ensures access to powerful networks in the host societies (Pujolar, 2007c). These linguistic behaviours, once again, also go against the ongoing goal of Catalan normalisation and of the defence of a generalised, de-ethnicised and unclassed public social use of the language, particularly in neighbourhoods where Catalan was (and still is) very marginal, more specifically in transnational discursive spaces like *locutorios*, where the use of Catalan is the marked exception.

The generalised practice of accommodating non-migrants in Spanish, together with this institutional discourse which, by contrast, makes every attempt to 'de-authenticise' and foster it as the public post-national language of social cohesion and 'integration', may be sending transnational populations ambivalent messages about the sociolinguistic landscape of Catalonia (Pujolar, 2010). This may partly explain why some migrant informants show a lack of awareness of, and lack of interest in, or a willingness to detach themselves from, the local political conflicts and the language-triggered identity-related issues in Catalonia (see also Newman *et al.*, 2013: 201; Pavez Soto, 2011: 72; Trenchs-Parera & Newman, 2009: 517).

It has frequently been argued that the groups of migrants who generally show more conflictive, negative and even patronising attitudes towards the Catalan language are the Latin Americans (Corona *et al.*, 2013: 182; Garzón, 2012: 2498, 2507; Marshall, 2006: 173; Marshall, 2007; Newman *et al.*, 2013; Pavez Soto, 2011: 81; Trenchs-Parera & Newman, 2009: 510), who, unlike migrant populations of Pakistani or Moroccan origin, for instance, are in the unique situation of being able to claim that their language is also official in Catalonia and, therefore, as valid and legitimate as Catalan (Alarcón & Garzón, 2011a: 148). This is especially so among Latin American people of middle-class, educated background whose migratory trajectories led them to occupy positions as unskilled workers in peripheral Catalan areas, where their credentials and their former social standing are not recognised. These experiences are frequently lived as a difficult downward socioeconomic step, which in part is expressed as a rejection of Catalan – and a rejection of what they perceive as institutional laws enforcing a 'Catalan first' policy in Catalonia (Alarcón & Garzón, 2011c: 134; Corona et al., 2013: 183; Garzón, 2011: 46, 51; Garzón, 2012: 2500; Marshall, 2007: 155; Pavez Soto, 2011: 85; Trenchs-Parera & Newman, 2009: 518; Unamuno & Patiño, 2009).

The learning of the Catalan language on the part of Latin American populations may also be nuanced by their conception of Catalonia as a momentary crossing point in their prospective transnational journeys, or by the desire to return to their home countries in the near future, which may render the investment in it less necessary (informant James, for instance, was not sure about staying in Barcelona). Besides, the Latin American migrants' identity work through the non-choice of Catalan may reveal not a mere dismissal of the Catalan language, but a complex way of keeping their particular Spanishes alive and of maintaining their cultural distinction by publicly rejecting 'assimilation' to the Peninsular Spanish language (Marshall, 2006: 521; Trenchs-Parera & Newman, 2009: 509, 516–517).

Of course, during the fieldwork I also found that those Latin American and non-Latin American migrants immersed in the local economy of the Barcelona area conducted multilingual practices which included the routinised use of Catalan and of Catalan–Spanish bilingualism. This was so even for those who tended to present themselves as non-Catalan-speaking *locutorio* clients. For instance, I recorded that most users had inserted the Catalan farewell *adéu* (or, if shortened, *déu*) instead of the Spanish *adiós* into their routine talk in translinguistic Spanish, even when they saw themselves as Spanish monolingual speakers. Ronny, a middle-aged unemployed tailor born in Bolivia who came to El Paso on his own in 2005, illustrates these claims in Excerpt 5.5.

Excerpt 5.5. The use of Catalan by migrants who present themselves as Spanish monolinguals

@Location: 1 September 2008. Bar near a *locutorio* in El Paso. Vallès Occidental.

@Bck: The researcher (RES) asks Ronny (RON) how many languages he speaks. He first presents himself as a Spanish monolingual speaker, but he then uses Catalan, too.

01 *RES: lo que me interesaba para el trabajo es -, a ver <cuántas lenguas hablas> [¿].
%tra: what I'm interested in in this study is -, let's see <how many languages do you speak> [¿].
02 *RON: eh yo soy unilenguado [=! laugh].
%tra: eh I am one-languaged [=! laugh].
03 *RES: sí¿
%tra: really¿
04 *RON: no # mentira <es un punto y aparte> [//] entre comillas.
%tra: no # [that's a] lie <it's something put aside> [//] in inverted commas.
05 *RES: qué +...
%tra: what +...
06 *RON: + ^ a ver un poco de: entiendo catalán pero <no> [/] no molt bé però +...
%tra: + ^ let's see a bit o:f I understand Catalan but <not> [/] not very well though +...

Ronny presents himself as an *unilenguado* (literally, 'one-languaged', that is, monolingual) in line 2. However, he then adds that this is not really true because he also understands Catalan. In fact, he ends up code-switching to Catalan in line 6. For transnational informants like Ronny, then, the Catalan language has become part of their linguistic repertoire, but its use is frequently framed within a Spanish monolingual ideology. This excerpt may be illustrative of the migrants' other-ascribed and self-attributed 'bad' or 'incomplete' competences in Catalan, which is highly counterproductive, for it further boosts the reluctance to use this language in public. Consequently, some of the migrants who state that they have not incorporated Catalan into their linguistic repertoires may have in fact achieved a certain command of Catalan–Spanish bilingualism, despite their claims to the opposite.

Figure 5.3 shows an entry in Shabbir's notebook which illustrates some of these normalised Spanish–Catalan bilingual practices by non-Latin American transnational migrants, in this case in the written mode. When I photographed it in the *locutorio*, Urdu-dominant Shabbir wanted to become a taxi driver, which requires a knowledge of Catalan in Barcelona, and so he had started to learn written Catalan (as well as Spanish), sporadically

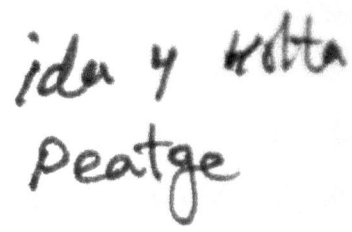

Figure 5.3 Bilingual written practices in Catalan and Spanish. Shabbir's notebook. *Locutorio* in El Paso. Picture taken by author, 27 July 2008

attending some of the Catalan lessons provided for free in a town nearby (he was the only informant who attended Catalan language courses). The words with which he practised include: 'round-trip' (*ida y vuelta* in Spanish; but here *ida y volta*, where *volta* could be either the 'misspelled' Spanish *vuelta* or a literal translation of Spanish in Catalan which results into non-standard *volta*) and toll (*peatge* in Catalan).

Finally, I encountered yet another situation with respect to the use of Catalan among migrant populations in El Paso. I found some migrants who had fully integrated the Catalan language into their linguistic repertoire, despite the fact that they did not normally employ it in their *locutorio* networks. This was the case of the *locutorio* worker who occupied Naeem's place by the end of the fieldwork, Aaliyah from Morocco, who had accessed Catalan via her previous job as a waitress in the tourist Catalan coasts (she had a Catalan-speaking superior there), and of yet another Moroccan *locutorio* worker in the same neighbourhood, Amal, who perfectly commanded the language, as I discovered only after first addressing him in this language.

Non-elite allochthonous minority languages

The migrants' non-elite allochthonous minority languages coexist in the *locutorio* on a daily basis, and there seems to be an implicit agreement that that is their place. *Outside* these migrant-regulated spaces, many transnational migrants make a list of the languages expected by the 'whites' when they are directly asked about their linguistic repertoires (Díaz, 2004). Thus, in the lists of languages that most migrants provide to their neighbours (as well as to researchers), for instance, they tend to place the Spanish language first; then they add English and/or French as the powerful linguae francae; and, finally, they make a general brief comment about their other languages – and, in the Catalan context, they may add Catalan, last of all. Inside the *locutorio*, though, this order seems to be reversed, as illustrated in the list of languages provided by Abou from Guinea in our interview, presented in Excerpt 5.6.

Excerpt 5.6. Presenting the migrants' language repertoires inside the locutorio

@Location: 1 September 2008. Bar near a *locutorio* in El Paso. Vallès Occidental.
@Bck: Abou (ABO) tells the researcher (RES) about his linguistic capitals, making a list of the languages that he feels he commands.

```
   01 *RES:   muy bien y <cuántas lenguas habla> [¿].
      %tra:   ok and <how many languages do you speak> [¿].
→  02 *ABO:   yo habla mi dialecto y habla francés y inglés un poco de cast
              +...
      %tra:   I speaks my dialect and speaks French and English a bit of
              Cast +...
   03 *RES:   + ^ el castellano <perfecto> [>] lo hablas.
      %tra:   + ^ Castilian <perfect> [>] you speak it.
→  04 *ABO:                               <castellano> [<].
      %tra:                                <Castilian> [<].
   05 *RES:   vale.
      %tra:   ok.
```

When I ask Abou about his repertoire of languages, the linguistic hierarchies he presents to me place his *dialecto* ('dialect')[55] first, followed by French, the colonial language in Guinea, then by English and, finally, by the taken-for-granted language of integration that he had mentioned before: Spanish (lines 2 and 4). He does not name the 'dialect' (he did, later in the interview; that is how I learnt that he spoke Pulaar, Susu and Manya), perhaps assuming that I would not be interested in them, or that I would not know of them. At least, though, in this particular discursive space the frequently backgrounded non-elite languages came first, before the dominant ones.

The visual iconography of *locutorios* also shows that both users and workers may at times place their allochthonous languages in a visible, prominent position, possibly as a way to symbolically attribute to them the degree of recognition that they tend not to have outside migrant-regulated institutions in Catalonia, and to compensate for their smaller presence in

Figure 5.4 The symbolic place of non-elite allochthonous codes in *locutorios*. Unconventional 'no smoking' sign handwritten in Arabic and Spanish. *Locutorio* in El Paso. Picture taken by author, 15 October 2008

the oral mode, in public. This is illustrated in Figure 5.4, which shows a handmade 'no smoking' sign first written in Standard Arabic in the Arabic script and then in Spanish, in the Roman alphabet. The Arabic part of this sign literally reads 'It is forbidden to smoke and thank you', and includes the diacritical marks usually employed in Moroccan primary schools to indicate vowel length. The part written in unconventional Spanish, below, says 'No smoking' (it is unconventional because the standard form of *proibido* is *prohibido*).

This Arabic–Spanish notice was shaded with a pink highlighter by an adolescent of Moroccan origin on her own initiative, with Naeem's consent. Of course, a printed, official 'no smoking' notice was hung at the *locutorio* entrance (as is compulsory in public businesses in Catalonia), so the new sign apparently provided unnecessary information. However, this young social player, who seemed to be willing to present herself as a cosmopolitan multiliterate teenager, decided to put hers up in order to give Arabic a symbolic public place, but without discrediting the actual translinguistic lingua franca of the institution, which she showed that she also commanded: Spanish.

As outlined in the introduction to this chapter, this acknowledgement of linguistic diversity does not mean that migrants in their self-regulated institutions give free rein to their non-elite allochthonous codes. The types of non-elite multilingualisms that I encountered there (instances of which have been presented in Figures 5.2 and 5.3) are actually far from liberating, in the sense that they impose a Spanish-unified floor and they strongly *regiment* the migrants' multilingual behaviours by silencing some of their linguistic capitals, which exacerbates their linguistic marginalisation, in their own alternative spaces, in inter-group interactions. In fact, I found that non-elite allochthonous languages publicly compete for a place in the *locutorios'* sociolinguistic hierarchy, as exemplified by the incidents presented in Snapshot 5.4, taken from my field notes from 22 August 2008.

Snapshot 5.4. The unseen competition over linguistic capitals among rival migrant groups

When Naeem has not been paid for weeks, or when he is exhausted and fed up with his clients' complaints, a series of ethnolinguistic labels are immediately put into motion. With so many people, today's afternoon has not been particularly good for him. 'Aquí hay mucho árabe, una mierda, hablan mal' ('Here there are many "Arabs" (Moroccans), it sucks, they speak badly'), he complains. Shabbir agrees, giving a disapproving look at little Yalda, a 12-year-old clever girl of Moroccan background who, extremely bored, plays near the counter. Shabbir agrees with Naeem's ideas concerning the Moroccans' linguistic practices, but he then adds that the rudest speakers here in El Paso are 'las españolas' (the local 'Spanish women'), whom he sees as impolite and as not having any manners: 'dicen cago en Dios, coño, joder... ¡Esto es muy malo, eh!' ('[They] say goddamnit, damn, fuck.... This is really very bad, eh!'). Then outspoken Yalda, still waiting for her older brother to finish playing on the computer, takes her opportunity to categorise back and jokingly tells him: 'pues a mi no me gusta el pakistaní, sólo el árabe' ('I don't like "Pakistani", I only like Arabic').

As a reaction to these kinds of verbal duels, these migrants, who themselves navigate through a multiplicity of multilingual and multimodal language practices, turn their non-elite allochthonous codes into *hidden-in-public languages*. The claiming of a place in the hierarchy, then, is conducted somehow silently, for example through the colonisation of the *locutorio*'s wall, which, in fact, provides the best iconographic representation of the

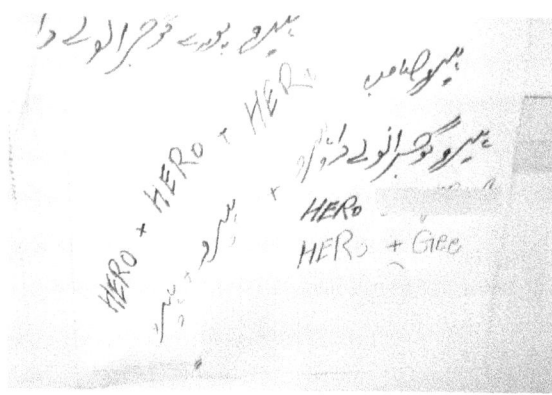

Figure 5.5 Hidden-in-public allochthonous codes in *locutorios*. *Locutorio* of El Paso. Picture taken by author (with selected details removed), 1 July 2008

linguistic landscape of the place. An example of this is provided in Figure 5.5, which shows a notice in Panjabi – surrounded by the word 'hero' (equally used in English and in Panjabi when employing the Roman alphabet) – calling persons from Gujranwala, in the Punjab area, heroes: 'You are the hero of the whole Gujranwala', or 'Hero puray Gujranwale da' in a Panjabi version written in the Roman alphabet.

This notice was written by a young man of Pakistani origin as a way to publicly fight against his group's shared code's 'ideological erasure' (Irvine & Gal, 2000) in El Paso, and to show affinity to and solidarity with the transnational local Pakistani network, members of which were frequently attacked (verbally) by other *locutorio* users (the message was actually directed to Naeem, the *locutorio* worker, who was in fact born in that very city). This notice is also the Pakistani network's way to symbolically claim belonging to the Catalan urban floor, as well as to the *locutorio*, and to more generally challenge the established institutional orders which dictate that the public language there shall be Spanish, and Spanish only.

The 'Everyone's Spanish' Paradox: Subversion and Self-Discipline in Prevailing Linguistic Regimes

In the previous section, I have shown that the alternative sociolinguistic hierarchies regulated by highly heterogeneous groups of migrants foster the regimentation of clients into a public Spanish-unified floor, but I have not yet gone into much detail about the type of translinguistic Spanish circulated inside *locutorios*, and I have not analysed its social meanings in depth either. Since, due to confidentiality constraints, I did not record naturally occurring interactions inside *locutorios* (see Chapter 1), I here focus on written data; in particular, on the migrants' new literacy and numeracy practices, in order to explore their use of this alternative code, which I have called 'everyone's Spanish' (Sabaté i Dalmau, 2013b: 122). More specifically, I analyse two different text messages, four handwritten room-for-rent advertisements, and a poster with a discount plan printed in the local lingua franca, transnational Spanish, all written by migrants who had not been schooled in Spanish and who made use of a diversity of communicative frames and grassroots literacy practices denoting multiple (self-)learning contexts linked to the use of ICT, which have frequently been neglected in migration studies (Androutsopoulos, 2006b: 522; Jacquemet, 2010: 62).

I approach these data from a user-centred perspective, in order to provide an understanding of those linguistic features relevant to and meaningful *for* the informants themselves, avoiding *a priori* assumptions or taken-for-granted preconceptions about their social uses and indexicalities in Catalan urban neighbourhoods (following Álvarez-Cáccamo, 1998; Auer, 1998; Danet & Herring, 2007).

SMS 1, sent on 29 July 2008, to Jenny (of Spanish descent), by her partner, Mohammed ('Moha'), a blue-collar employee born in Morocco working in an electrical appliances factory, provides the first insights into the migrants' everyday literacies in translinguistic Spanish.

SMS 1. 'Everyone's Spanish' (I)

'Ola carinyo estue mue continto para verte corazon estamos jontos y na braso' (Original SMS reproduced verbatim)

'Hola cariño, estoy muy contento de verte, corazón. Estamos juntos. Un abrazo.' (Transcription in Standard Peninsular Spanish, my version)

'Hi love, I'm very happy to see you, sweetheart. We are together. A hug.' (A possible translation into Standard English, my version)

Mapped upon the Spanish language, Moha's intimate text message shows some traces of vernacular or grassroots literacy, as well as of a 'low' command of the conventional graphic representations of Standard

Peninsular Spanish. It also includes some *heterography*, or 'the deployment of graphic symbols that defy orthographic norms' (Blommaert, 2008: 7), and some traces of xenoglossy as well, understood as the linguistic ability to access and to appropriate a code without having fully understood or learnt it (Jacquemet, 2005; Jørgensen *et al.*, 2012). This occurs, for instance, when he writes 'na braso' (arguably equivalent to 'an arm' in some Spanish accents), possibly a morphological reinterpretation of 'abrazo' ('hug'), because both the standard and the unconventional forms (that is, both 'un abrazo' and 'na braso') are phonetically equivalent in many Spanishes, resulting in some 'truncations' (Anis, 2007: 90). Truncations are groups of words where some letters are kept separate from each other, rather than joined together, as happens, generally, in the Arabic script (Palfreyman & Al Khalil, 2007: 49).

Moha's spelling difficulties and orthographic uncertainty may speak of his multiple literacies, since he also makes use of some transcodic marks from Arabic phonology that may reflect 'accent', as in the words *estue* ('I was'), *continto* ('happy') and *mue* ('very'). Thus, his SMS also contains a fair amount of orality traits, as he is himself immersed in the current 'post-literate technological context' (Pennycook, 2007: 13), which allows for meaning-making across oral and written modes.

At the same time, Moha shows a unique command of the type of language employed for in-group membership, proximity and intimacy, as he semiotically transmits the aesthetic flavour of the spoken word (that is, his *voice*) to Jenny. He does so in more horizontal, technology-mediated 'non-official ways of speaking' (Thurlow & Mroczek, 2011: xxiii), for example through the absence of punctuation and orthographic conventions (like accents or commas) and through word forms like *ola* (instead of *hola*), which, he claims, he shortened on purpose (I read and interpreted these SMSs with the informants). Thus, this written alternative language is also a 'symbol of commitment' to his network (Androutsopoulos, 2007: 349) which includes (un-)conscious use of anti-normativity and iconicity, and which is part of the non-standard 'textese' variety or the street 'we-code' circulated among the migrant and local Spanish-using members of one of his important social networks in El Paso.

Shabbir, the 41-year-old construction worker from Kashmir, provides more insights into this sort of entextualised multimodal Spanish in SMS 2, which is a Christmas greeting that he sent me on 24 December 2008 (also reproduced with informed consent).

SMS 2: 'Everyone's Spanish' (II)

'Hola.buens noche . srts maria mi Hermana felis novides.from m. Shabbir mughal' (original SMS reproduced verbatim)

'Hola. Buenas noches, srta. Maria, mi Hermana. Feliz Navidad. De M. Shabbir Mughal.' (Transcription in Standard Peninsular Spanish, my version)

'Hello. Good night, Miss Maria, my sister. Merry Christmas. From M. Shabbir Mughal.' (A possible translation into Standard English, my version)

Shabbir's text message also includes non-elite multilingual practices in Spanish and in English, and the use of phonetic spelling, too (as seen in *felis* for *feliz*, which means 'happy'; or as in *novides*, for *Navidad* or 'Christmas').[56] SMS 2 bears traces of xenoglossy, too, in this case involving morphological aspects that concern adjective–noun agreement in Spanish, as in the unconventionally inflected farewell *buens noche* (instead of *buenas noches*).

Shabbir's discourse differs from Moha's in that it is carefully planned and organised. It starts with two greetings and with a very polite term of address (*srts*, for *srta.* or 'Miss'), followed by a Christmas greeting and, finally, by an accurate farewell and by the name of the SMS sender, Shabbir. Moreover, it is carefully punctuated, with stops dividing up each discursive unit or speech act. He even employs the uppercase form for the familiar term of address *Hermana* ('sister'), perhaps as a marker of deference.

These two SMSs demonstrate that, through ICT, transnational migrants may take up and 'do' many different genres and registers (Pennycook, 2012: 98), developing a command of various degrees of formality and a very wide range of politeness styles, from very informal – as in the case of Moha – to (hyper-)polite – as in the case of Shabbir (for more examples of such SMSs, see Sabaté i Dalmau, 2012c). Shabbir refuses to use informal contractions (he typed the full form of *hola* for 'hello', even though he said that he knew about the frequent use of *ola* in the neighbourhood), and he also includes a face-saving term of reference which indexes social distance (*srts*). Besides, he marks the closing discursive unit by switching from Spanish to English, the language of prestige among migrants of Pakistani origin that he also uses in his offline linguistic practices as a marker of a high socioeconomic class and educational distinction. The fact that he emphasises his 'Mughalness' in his SMS (which simultaneously indexes Muslimness, wealth and power; Farlex, 2012) provides further evidence for this claim. Thus, the switch from Spanish to English may be 'not just a contrast of languages but rather language styles selected for different kinds of identity work' (Androutsopoulos, 2011: 294), in this case, that of ascribing to himself a serious and respectful technoliterate Muslim identity.

Normally, this migrants' written translinguistic Spanish is socially devalued as being indexical of 'literacy gaps' and of 'faulty interlanguage stages', even by people of their own networks, such as by the first SMS receiver, Jenny, who, when talking about Moha's text message, said 'tendré que enseñarle a escribir' ('I'll have to teach him how to write'). This is so because non-standard representations of language, which 'have the potential to challenge linguistic hierarchies' and 'to make non-standard voices visible/audible in a medium that habitually does not recognize them', are always framed within what is considered to be the standard, legitimate norm (Jaffe, 2000: 498). Thus, to mainstream society, as well as to those who have been schooled in the Spanish language, Moha's and Shabbir's text messages are written in a *not-quite-language* (Gal, 2006: 15), employed by users who are constructed as 'deficient', 'deviant' or 'incompetent' (in Gumperz's sense; Gumperz, 1986), in this case in written Spanish. In fact, this ecumenical code with transcodic marks has derogatorily been called *immigrantese*, a label which, following Ferguson's (1975) analysis of 'foreigner talk', infantilises and pejoratively simplifies the migrants' wide range of oral and written linguistic capitals in the host societies' languages.[57]

However, all sorts of notices in this hybrid 'agrammatical' language and this 'mixture' of codes and overlapping communicative frames have colonised the wall of the *locutorio* and have, indeed, become the migrants' translinguistic fully fledged 'everyone's Spanish', employed for local navigation and reterritorialisation. This demonstrates that migrants have been able to appropriate and transform the standard norms of Spanish as well as the established literacy regimes of the host societies in a bottom-up manner. To start with, this alternative polyphonic lingua franca has allowed them to make available the local information which is vital for getting organised around the neighbourhood across diverse migrant groups, that is, for finding a commonly shared 'linguistic way of (inter)acting' (Thurlow & Mroczek, 2011: xxx) in El Paso, for example to find a room for rent, as I will now illustrate.

Figure 5.6 shows two room-for-rent advertisements written by the same informant, a neighbour from Morocco, on two different days. Like SMSs 1 and 2, the advertisements show traces of xenoglossy, as illustrated by the fusion of impersonal Spanish constructions (starting with *se*) with the simultaneous 'misuse' of the first person singular inflection (*alquilo* in Figure 5.6a) or with the use of unorthodox ('faulty') word order (the

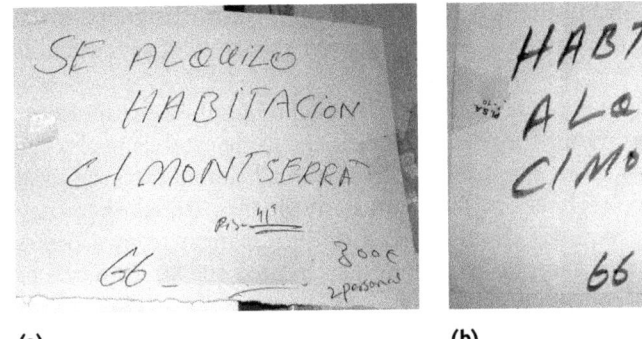 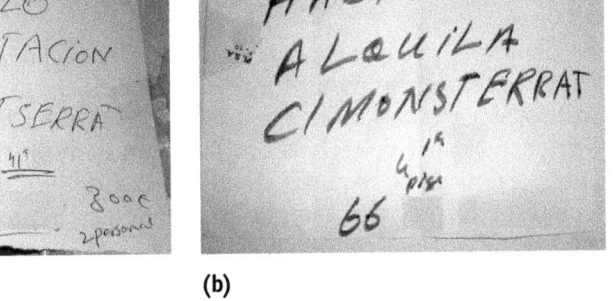

(a)   (b)

Figure 5.6 The social uses and meanings of 'everyone's Spanish'. Room-for-rent advertisements posted by the same user, (a) on 9 September and (b) on 16 September 2008. *Locutorio* in El Paso. Pictures taken by author (with selected details removed)

direct object is ungrammatically followed by the verb in Figure 5.6b). The two advertisements also include translinguistic marks, where *montserrat* (a Catalan street name) coexists with *monsterrat* and where *habitacion* (meaning 'room') coexists with *habtacion* – other variations of the word 'room' that I photographed included *abitasion* and *habitasion*. In the case of *monsterrat*, the *locutorio* informants stated that they remembered the name of the street due to its similarity with the English term 'monster'. *Habtacion* (instead of *habitación*) could be an involuntary 'error', since the author was familiarised with the Arabic alphabet, which would explain the absence of vowels (the Arabic script includes 28 letters, which usually stand for consonants only). Both *monsterrat* and *habtacion* are thus transidiomatic marks indexing multiple (Arabic, English and Spanish) literacies.

The flat number in the two pictures also has two modalities: 41ª (flat 41st in Figure 5.6a) and 4ª 1ª (4th floor, 1st door, Figure 5.6b). This, though, does not seem to be a problem when interpreting them, since both advertisements prioritise another type of information. Despite the fact that the street name is 'wrongly' written and that the flat number is unclear, what is relevant here (apart from the price) is that they indicate that there is a room available, and that this is located in a specific area, for example near the public transport line (I also interpreted several advertisements with the migrant *locutorio* users who wrote and who read them). This, in turn, indicates that this written code is to be transtextualised (that is, decoded) within the *immediate social context* in which it gets circulated, which means that these literacy and numeracy practices in the Spanish language make sense and become relevant only within migrant-regulated spaces such as *locutorios*.

Through these literacy practices, transnational migrants are redistributing resources such as rooms for rent by playing the game established by the host societies, but without the need to be fully enculturated into the local written normativity regimes. In fact, some informants stated that they simply calqued the required chunk of language in Spanish from a peer's room-for-rent model or sample, and then just changed the street name or the street number accordingly.

Through their aesthetics the advertisements also suggest that these rooms are rented out by a migrant person who frequents the *locutorio* networks in the neighbourhood – perhaps someone with similar transnational experiences. Nicely printed advertisements in Standard Peninsular Spanish, by contrast, did not transmit the style of a non-elite multilingual 'voice' and tended to be attributed to outsiders. They in fact indexed mistrust and, at times, even some sort of 'poshness', which may explain why they were not successful there and quickly disappeared from the *locutorio* wall. Thus, this code is mobilised to protect and distribute the resources of the Catalan urban floor among in-group migrant members only, in their own self-regulated institutions, outside mainstream realms.

These multilingual practices mapped upon translinguistic Spanish are so frequently used that they have entered the circuits of press capitalism, to the extent that they have even become the 'alternative' corporate language of some 'ethnic' businesses targeting migrants, like *locutorios*. This is shown in the advertisement for the international phone card Siempre Latina (with its discount plans for July 2008), presented in Figure 5.7.

This poster, which was brought by the Urdu-dominant distributor of Pakistani origin who provided the call cards in El Paso, shows traces of grassroots literacy and heterography mapped upon a predominantly Spanish-only landscape. The non-standard 'misspellings' and the neographical transformations of some geographies of the globe (highlighted in Figure 5.7) include: *Asunclon* (for Asunción); *Bagota* (for Bogotá); *Brazil–Brasil*; *Ukraine–Ukriaine*; and *Dominica* (both for Dominica and Dominican). The advertisement also includes transidiomatic marks (as in the use of the English word 'mobile') and chunks of commercial formulaic talk in Spanish, as in *atención al cliente* ('customer services').

Figure 5.7 suggests that this 'everyone's Spanish', which is not found in the advertisements issued by big multinationals, has somehow been fetishised and employed to do business in a more horizontal way, from migrant hands to migrant hands. I do not claim that we are witnessing a

Figure 5.7 'Everyone's Spanish' in migrant-tailored corporate information. Discount plans by Siempre Latina. *Locutorio* in El Paso. Picture taken by author (with selected details removed), 31 August 2008. (Highlighted details discussed in the text)

real commodification of transnational grassroots literacy, heterography and xenoglossy, but these linguistic features are already gradually becoming part and parcel of the new migrant entrepreneurs' repertoire of assets to target a migrant clientele. These linguistic resources are mobilised from within, with an in-depth 'insider' knowledge of the migrants' real multilingual capitals, which, again, shows that the globalised new economy is moving towards its *migrantisation*, at least in what is considered a 'peripheral' part of the ICT business world.

Overall, these practices demonstrate that migrants have the capacity to autonomously insert themselves to (and to appropriate) hegemonic contemporary communicative and literacy regimes in order to resist their linguistic marginalisation. In this sense, these devalued vernacular practices are largely *subversive* of a dominant cultural order, because they are non-normative and counter-hegemonic, and because, by their very hybrid, translingual nature, they publicly 'propose and embody alternate models of the social world' (Gal, 2001: 425). Besides, they also transgress the deeply rooted ideological foundations of the 'native speaker' (Pennycook, 2007: 36): this 'Spanish' is, in theory, no longer monopolised by those who come from an 'authentic' Spanish-speaking background.

I do not claim, though, that this translinguistic code is an empowering or a democratised language for all migrant groups, for, as Jacquemet (2010: 63) warns us, the emerging heteroglossic codes of the global era 'are still inserted into a global indexical order which assigns superior values to certain systems of communication'. In the Catalan context, the flexibilised Spanish found in *locutorios* is clearly inserted into the shared regime of thought that I have analysed in the previous section, where non-Spanish dominant migrants show a tendency to follow, and to discipline themselves in, the Standard Spanish monolingual norm which prevails in mainstream society, discrediting and 'correcting' their own non-conventional functional written Spanish, as shown in Figure 5.8.

Figure 5.8 presents two room-for-rent advertisements for the same flat written on two different days by the same Moroccan *locutorio* user. With a non-inflected bare infinitive, they both read: 'rent room', in non-standard Spanish. The difference between the first advertisement (a) and the second (b) is that the latter has undergone a 'transposition' (Blommaert & Rampton, 2011: 10), that is, a revaluation whereby the letters have been embellished through elongation and have been highlighted in yellow in order to give them more prominence and visibility on the *locutorio*'s wall. Besides, the word 'room' has been corrected to meet the dominant normativity standards, so that in the second picture (b) the first grapheme, '*a*', has become an '*h*', and a little vowel, '*a*', has been inserted on top right after it, resulting in

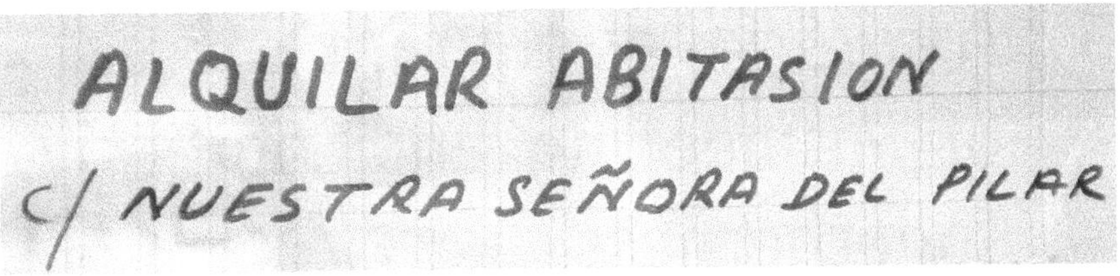

> ALQUILAR ABITASION
> C/ NUESTRA SEÑORA DEL PILAR

(a)

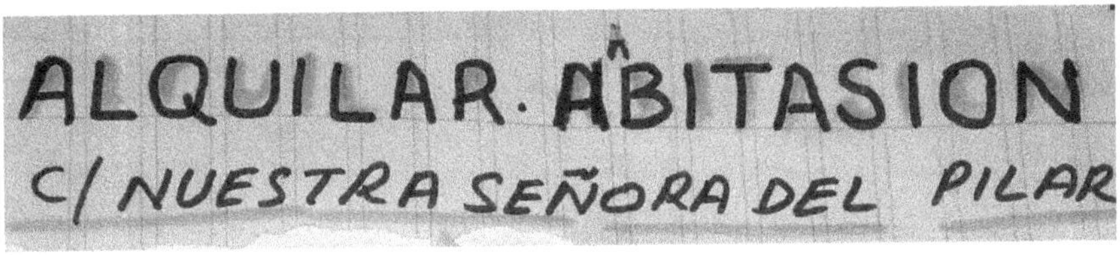

> ALQUILAR· ÂBITASION
> C/ NUESTRA SEÑORA DEL PILAR

(b)

Figure 5.8 Self-disciplining practices in written Standard Spanish. Two room-for-rent advertisements for the same flat, posted (a) 17 July and (b) 20 August 2008. *Locutorio* in El Paso. Pictures taken by author.

a 'respelling' (Tagg, 2012: 4) which better represents the more standard word *habitasion* (the word in Peninsular Spanish, though, is *habitación*). This bears witness to the Pakistanis' and the Moroccans' efforts to meet the standards of the lingua franca of the place and to be able to compete, on equal grounds, with those who are already enculturated into the regimes of normalcy of the host societies, particularly with the locals and with the Latin American schooled migrants, for example in the flat-sharing market. Their self-disciplining in Spanish is such that some have even *Spanish-ised* their names when introducing themselves to Spanish-speaking populations in El Paso. Sheema, for instance, calls himself Chema (/'tʃema/) and Shabbir, Xavi (/'tʃabɪ/, a Catalan name which is common among Spanish-born adults in El Paso).

In the following chapter, I complement this analysis of the ambivalences of the migrants' complex linguistic practices and ideologies with an in-depth exploration of their intra- and inter-group social categorisation rules and identity positionings, which will provide a more complete picture of the social structuration dynamics of heterogeneous migrant networks in their non-mainstream institutions.

6 *Locutorios* as Migrant Spaces of 'Mismeeting' and Conflictive Togetherness

> *The socialization of society – the construction of a shared cultural practice that allows individuals and social groups to live together (even in a conflictive togetherness) – takes place nowadays in the networked, digitized, interactive space of communication.*
> Manuel Castells (2004: 30)

As outlined in the previous chapter, migrant-governed transnational spaces of resistance and ICT-mediated subversive communication are not always empowering for all their users, despite the undeniable function they have as the real challengers to present-day exclusionary citizenship regimes. In this chapter, I argue that the tremendous social role that *locutorios* play should be problematized in order to fully understand not only the positive but also the negative effects that these 'ethnic' businesses may have upon diverse groups of migrants, particularly upon the transnational life trajectories and the work prospects of those at the bottom of the social ladder, in late capitalism.

As a window into the highly complex power dynamics that regulate socially hierarchized migrant networks in Catalonia, in this chapter I investigate the ways in which various *locutorio* social players mobilise not only their multilingual repertoires and literacy capitals but also their ethnic, gender, class, religious and educational identity attributes as well as their citizenship status in order to negotiate their belonging to, and their entrance into, these defensive transnational spaces. I do so through a unique analysis of the migrants' locally situated self-ascribed or hetero-attributed social categorisations, multiple acts of identity, multi-local affiliations and fluid presentations of the self, all of which allow for a complexification of their largely unknown intra- and inter-group 'othering' processes and practices.[58]

I explain that some of these discursively constructed emergent migrant identities may involve pejorative social labels which are marginalising, such as *gorronas* (young female Latin American 'scroungers'), *ladrones* (male

Moroccan 'thieves'), *lagartas* (female 'Romanian lizards' or sexed reptiles) or the time and money 'wasters' (Pakistanis who make daily use of the internet), all of which reveal the ways in which diverse migrant groups strategically network side by side inside *locutorios*, but certainly not always in comradeship or solidarity. With this, I suggest that these alternative institutions of transnationalism may well turn into a disempowering 'togetherness of loners' (Bauman, 2001: 68) for some migrants, or into a collective space of 'mismeeting' (Larsen & Hviid Jacobsen, 2009: 83) which individual migrants inhabit with the aim of accessing a resourceful protective umbrella but at the cost of coping with conflict-ridden, compelled cooperation with hostile groups of migrant 'others'. I thereby argue that technologised institutions like the *locutorios* may be a double-edged sword, in the sense that they may hinge upon what Castells (2004: 30) calls 'conflictive togetherness', that is, upon multi-valued, ever-changing social relationships based on simultaneous trust and distrust, mutual friendship and enmity, love and competition, and neighbourliness or comradeship and, at the same time, malfeasance, hatred and even violence or harassment (see also Lewicki *et al.*, 1998; Tilly, 2007).

In the second part of this chapter, I depart from the migrants' disparaging ethnolinguistic rules of social categorisation to focus exclusively on the most negative consequences of an extreme case of 'conflictive togetherness' established between the *locutorio* worker central to the present study, Naeem, and the mass of *locutorio* users, on the one hand, and between the *locutorio* worker and his Pakistani superior, on the other. More specifically, I explore the attributed workplace roles and identities that Naeem had to fight against when he was positioned by his newly empowered clientele as a privileged, wealthy, documented and technoliterate 'employer', while he, in reality, was an exhausted, pauperised worker undergoing all sorts of exploitative practices in that particular *migranticised* workplace realm of El Paso.

I choose to analyse Naeem's intricate identity management and identity display as a specific example of the subject positions of a group of transnational workers because this can help us understand the legitimate and non-legitimate habitable ethnolinguistic identities emerging in the postnational capitalist scenario in which we now live: the *language workers* of the tertiarised new economy.

We already know that, in the current formal circuits of commerce, language industries (call centres, tourist agencies and so on) target their clients by trapping workers into marginalising linguistic regimes and into imposed presentations of the self. These include the employees' tailorisation of talk in standardised interactions in given legitimate languages or accents and the enforced enactment of particular 'authentic' identities, under close surveillance and tremendous work pressure (Boutet, 2006, 2008, 2012;

Cameron, 2000; Da Silva *et al.*, 2007; Duchêne, 2009, 2011; Heller, 2003; Heller & Boutet, 2006; Roy, 2003; Sonntag, 2005; Taylor & Bain, 2005).

By analysing the *locutorio* workers' extremely hard workplace conditions, I try to gain a better understanding of what happens when the surveillance of adequate 'customer services communicative skills' is actually informally carried out face to face and in a bottom-up manner by highly demanding groups of clients who collectively project their fight against social stagnation at the *locutorio*'s front line, addressing insulting categorisations such as *nenaza* ('sissy'), *maricón* ('faggot') or *Moro de Al Qaeda* ('Moroccan Al Qaeda terrorist') to abused *locutorio* workers, who, alone, find themselves seeking to protect their transnational family's income and their visas through the deployment of reactive communicative weapons and impressively useful identity armours.

Migrant Identities and Power Dynamics in Non-Mainstream Worlds

A lot has been written on the types of ethnolinguistically driven social categories attributed to migrants by mainstream society in the Catalan context, and on the consequences that these may have for the different transnational populations who live in this part of the world. For example, it has been shown that these mobile groups more often than not tend to be categorised in many different institutional settings as inferior, disempowered, problematic or unreliable 'others' with 'deviant' linguistic capitals (Codó, 2008; Codó & Garrido, 2010; Moreras, 2007; Moyer, 2010, 2011; Pujolar, 2007b; Serra del Pozo, 2008). However, we still do not know much about what Vertovec (2001: 577) calls the migrants' 'internal identity dialectics'. We do not have much knowledge either of how migrants conceive of themselves, and of how they are characterised by other migrants when they mobilise their identity resources, their public social affiliations and their strategies of belonging (or non-belonging) in their own self-governed alternative spaces. It is because of their non-institutional, migrant-governed nature that *locutorios* allow for the unique observation of struggles for survival in contexts of precariousness, where what is at stake in the redistribution of social networking capital via identity negotiations often is one's transnational bread (Sabaté i Dalmau, 2012a).

The room-for-rent advertisements posted on the *locutorios*' walls may be a window into the migrants' publicly displayed transnational local ties and identity affiliations, for they reflect the increased 'diversification of diversity' (Vertovec, 2007: 1024) of migrant groups in Catalonia, as well as

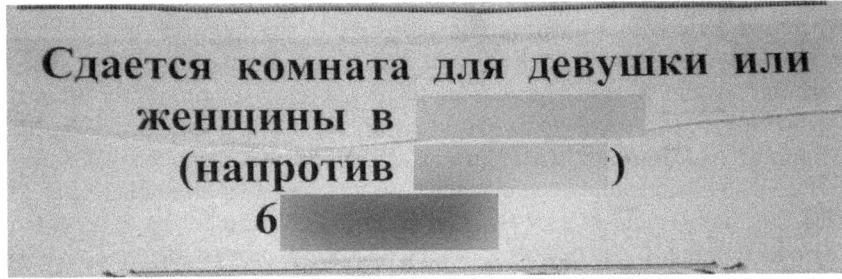

(a)

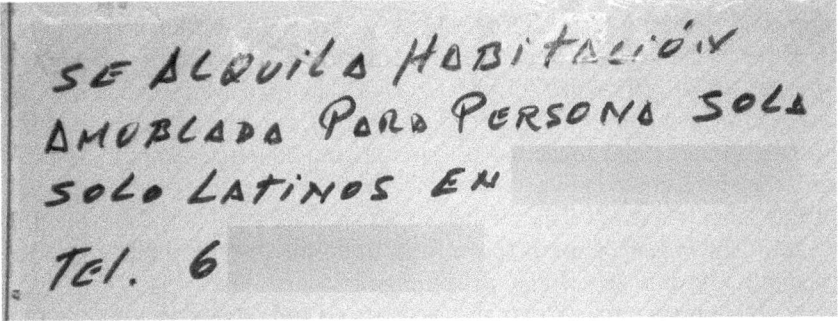

(b)

Figure 6.1 The migrants' social forms of differentiation in language and in identity display. Room-for-rent advertisements posted on the walls of the *locutorio* in El Paso (a) 19 September and (b) 1 July 2008. Pictures taken by author (with selected details removed)

the substantial degree of clienthood specialisation increasingly observable in their self-regulated businesses. These advertisements contain a series of linguistic features and social labels which have a clear demarcating effect and which, by suggesting the ethnolinguistic identities of potential users, embody the migrants' elaborate social forms of differentiation and of competition. Figure 6.1 shows such two advertisements which illustrate these claims. The first (6.1a) is written in Russian and reads 'Room for rent to a young woman, near X', and the second (6.1b) is written in Spanish and states 'Room for rent to a single person. Only Latinos'.

The advertisement in Figure 6.1a targets the sub-group of Russian-speaking and Cyrillic-reading women (including some of the so-called

Rumanas) for whom this code has become a small lingua franca in the neighbourhood. I argue that the advertisers resorted to this linguistic practice as a way to gatekeep this particular resource in an intra-group manner, possibly excluding all men, and those who were not ethnolinguistically aligned with the Russian language or who had no access to Cyrillic. In fact, the Moroccan, Pakistani and Latin American groups of *locutorio* users whom I followed also interpreted this specific choice of alphabet and of language in this way, so they did not see themselves as possible targets, since for them the advertisement was neither transparent nor fully decodable.

The advertisement in Figure 6.1b was written by a person from Latin America who was explicitly looking for a *'Latino'* flatmate. When asked, the auto-categorised *'Latinos'* or *'Sudamericanos'* (similar advertisements employed the latter term) stated that they used these social labels as a way to distance themselves from rival migrant groups who were not born into Spanish-speaking families – basically, Pakistanis and Moroccans, who, as has been shown above (in Chapter 5, Figure 5.8), also had a command of the Roman alphabet and made use of handwritten advertisements in Spanish and, therefore, could emerge as competitors in the local flat-sharing market. In this sense, these advertisements project the migrants' race to be the first to enter the *locutorio*'s marketplace, and to be the first to publicly inscribe their voice within the Catalan urban linguistic landscape.

These sorts of public 'internal' competitions and subject positions among migrant groups may be better captured through a fine-grained exploration of their locally situated social stratification practices, as illustrated in Snapshot 6.1, where I present several troublesome encounters involving: single men of sub-Saharan origin; young women from Eastern Europe and Latin America; and men and women from Pakistan, as well (I use fieldwork notes from 26 and 30 June, 9 and 22 July, 25 and 28 August, and 30 September 2008).

Snapshot 6.1. Hetero-attributed social identities: Migrants' locally situated social stratification practices

Very soon that summer I realised that the lowest rank in the social hierarchy of this particular *locutorio* is occupied by a small group of black migrants of sub-Saharan origin consisting of homeless men from Gambia, Mali, Nigeria and Guinea, who spend the night on the benches, and who partly make a living by selling pirated CDs and DVDs. In El Paso, they are systematically ostracised by the vast majority of *locutorio* clients and workers (regardless of their ethnolinguistic background), who keep interactions with them to the minimum. An instance of this was provided to me by Rafa, the adolescent born in town who always smokes at the *locutorio* entrance, when he, in a macho-like attitude, laughed at one of these men leaving the place by shouting (in Spanish): '¡A este lo pones dentro de un coche rojo y parece un Kit Kat!' ('Put this guy inside a red car and you've got yourself a Kit Kat!'), followed by hoots of mocking laughter from inside and from outside the *locutorio*.

The *'Rumanas'*, young women born in Romania (and, as I later discovered, also in Russia and in other regions of the former Soviet Union!) seem to occupy the next-to-lowest place in the *locutorio*'s hierarchy, for, on being systematically called *'lagartas'* (equivalent to 'lizards' but indexing something like sexed reptiles) by their rival Latin American young female social players, they are doubly stereotyped. They are presented not only as unreliable and as thieves but also as too sexually active, promiscuous. Lila, a young informal waitress from Uruguay, for example, insists that they are quick at emptying the young men's pockets, particularly those of the *'Pakistaníes'* like Naeem. She explains that they do so by bewitching them with their bodies, a 'trick' which includes the 'immoral' non-use of bras. Perhaps the boundaries between the *'Rumanas'* and the self-categorised *'Latinas'*, though, are mostly a matter of class or socioeconomic stance. The *'Rumanas'* in this neighbourhood, unlike the group of *'Latinas'* who institutionalised this *'lagarta'* pejorative label, have a flat to share to themselves. By contrast, the most visible group of very young women from Latin America survive by spending the nights in some of their friends' rooms (normally in Abde's flat) and even at their workplaces (such as Sheema's bar), where Lila spends most of her days. Once, one of these *'Rumanas'* approached the counter to ask for a €50 note to be changed, when Lila murmured: '¡Cuánto dinero, qué asco!' ('What a lot of money, how revolting!').

According to men of Pakistani origin like Shabbir, though, the *'Latinas'* are 'worse' than the *'Rumanas'* (they also employ these two social labels, in Spanish), because they always ask for all sorts of material goods for free, for which they are talked about as some sort of *'gorronas'*, or 'scroungers'. Naeem, the *locutorio* worker, complains that 'siempre piden llamar gratis, copas, siempre piden dinero' ('they always ask to call for free, for drinks, they always ask for money'). Sheema, the informal bar owner from Pakistan, agrees and adds: 'Mi madre tiene cáncer, mi hijo problemas, dame dinero' ('my mother has cancer, my son problems, give me money').

When asked about the women of Pakistani origin, though, they both claim that they do not even want to mingle with them. Sheema, married to a Pakistani woman, and with two adolescent children schooled in town, tells me that 'they do not speak', and Naeem adds that '¡Las chicas de mi país no saben nada!' ('the girls from my country know nothing!').

The identity dialectics described in Snapshot 6.1 show that, when fighting for a place and for a voice in the social hierarchies of the *locutorio*, migrants mobilise identity labels grounded on the imagined places of origin of particular transnational populations. They also show that they then link such perceived geographies with given personhood traits concerning gender, age or class, discursively fixing resourceful social categories such as the *Latinas*, the *Rumanas* and so on, routinely employed in the taken-for-granted language of the place, Spanish. These sort of categorisations (pejorative, in this case) are homogenising for whole groups or *collectivities*, although they emerge out of particular migrant individuals (like Lila) who act or who are acted upon as 'tokens', or as category representatives (Martín-Rojo, 2010). Snapshot 6.1 includes the discursive representations and the hetero-ascribed dismissive identity constructions of the following transnational migrant groups: the pauperised Africans dispossessed of some manly capital by being reduced to a 'chocolate'; the criminalised and perhaps masculinised 'dirty' Eastern European women; the abusive, partying Latin American women who have mastered the art of lying; and the 'language-less' (Blommaert *et al.*, 2005: 213) Pakistani 'dull girls' with a lack of 'Westernisation'.

Religious affiliation (or non-affiliation) plays an equally crucial role across transnational migrant groups. In El Paso, it is frequently Moroccan and Pakistani men who undertake rival religious identity enactment and identity construction practices in public when they strive to be the ones defining who counts as the legitimate local Muslims of the place. In Excerpt 6.1, informant Shabbir from Kashmir, the Muslim leader of the neighbourhood, for instance, claims that 'people from Morocco' steal women's bags in Barcelona (line 4), a practice which is decried as un-Muslim. He also makes it clear that he does not (want to) know about the Moroccan men's whereabouts in El Paso, and presents the Moroccans and the Pakistanis as two worlds apart, in turn positioning himself as a more morally righteous follower of the precepts of the Islam.

Excerpt 6.1. The discursive construction of 'Moroccan thieves': A competition over legitimate 'Muslimness'

@Location: 10 August 2008. Street near a *locutorio* in El Paso. Vallès Occidental.
@Bck: Prompted by the researcher (RES), Shabbir (SHA) talks about the disadvantages of modern metropolises like Barcelona City, which leads him to mention, and to disapprove of, people he calls 'Moroccan thieves'.

01 *RES: y qué es lo que más te gusta de www¿
   %tra: what do you like best in www¿
   %com: www is the town in Vallès Occidental where El Paso is located.
02 *SHA: www vives aquí gente tranquilamente y: yo quería vivir tranquilamente.
   %tra: in www lives here people peacefully a:nd I wanted to live peacefully.
03 *RES: claro.
   %tra: sure.
→ 04 *SHA: porque Barcelona hay mucho ruido y mucha gente # antes <sabes qué> [¿] gente de Marruecos porque no lo sé donde están robaron mochilas de mujeres y tal tal historias y no me gusta esto.
   %tra: because Barcelona a lot of noise and many people # before <you know what> [¿] people from Morocco because I don't know where they are stole bags from women and such and such stories and I don't like this.

This competition for the attribution of legitimate or non-legitimate Muslim capitals in non-mainstream realms passes unnoticed outside the *locutorio*, where the two groups (Moroccan and Pakistani) are normally homogenised and disparagingly regarded as a single mass of *moros*, another public discursive representation in Spanish (roughly equivalent to 'Arab', but with derogatory connotations) employed by non-foreign neighbours and, at times, by some migrant populations from Latin America.

Inside the *locutorio*, though, any relationship between Pakistani and Moroccan men can become violent over religious issues. Their internal disputes have become so tough that Naeem was once publicly harassed by a group of four '*moros*' for having eaten, drunk and smoked during Ramadan (he also uses the label *moro* inside the *locutorio*, but to contemptuously name only the local group of men from Morocco). And yet, simultaneously, in public, the two groups, in forced religious comradeship, have joined forces to petition the town council for the establishment of a mosque for all the Muslims in the town (this formal petition had not been successful at the time of writing).

In the analysis of the presentation of the self and of the 'other' in late capitalism, the artefacts of the globalised new economy, like mobile phones or music, understood as constitutive of the capitalist material cultures in which present-day migrant populations are fully engaged, also play a key role in social categorisation processes, normally in relation to the emergence of fluid and superficial 21st-century individual consumerist identities (Bauman, 2000, 2005; Giddens, 1991; Harvey, 2005; Pujolar, 2011; Slater & Miller, 2000; Vertovec, 2001). Thus, the ownership of ICT gadgets in general indexes specific gender, legality, educational or class identity attributes. Mobile phones, in particular, have become not only an integral part of the migrants' new modes of self-expression but also the recurrent metadiscursive terrain by means of which competition for transnational resources is played out and overtly negotiated, especially in their own *locutorio* phone shops, where all these consumer goods are available. In Snapshot 6.2 (which gathers field notes from 30 June, 22 July, 15 September and 8 October 2008), I exemplify some of these ICT-related identity performances, which include, for instance, the choice of handsets indexing a cosmopolitan female identity or a young masculinised hip-hop identity, all linked to manifold self- or other-ascribed identity traits, like 'poshness' or 'Spanishness'.

Snapshot 6.2. The migrants' fluid consumerist identities

In here migrants change their mobile phones very frequently. The more expensive and the newer the handsets are, the higher the social class they seem to index. I bring a leaflet from the multinational Orange with many handset pictures and ask informants which mobiles best serve the needs of transnational communication.

Lila, the waitress from Uruguay, tells us that she would really like to get the metal-pink Sony Ericsson Hello Kitty edition, which is the smallest handset with rounded shapes that seems to transmit the fashionable style of a cosmopolitan woman (it is advertised on television as such).

Merche's partner, also auto-categorised as a *'Latino'*, points to the latest Samsung model with internet connection that he got, in purple, but via Movistar, considered to be the most expensive company. A few days earlier, when he first showed it in front of everyone still inside its brand-new box, Naeem exclaimed that '¡cuesta más de 300 euros!' ('It costs more than €300!'). It seems that this particular mobile simultaneously indexes a high purchasing power, a considerable degree of technoliteracy in Spanish and a status of legality, because internet-endowed handsets are normally obtained via the formalisation of a contract. This may explain why Merche's partner is at times considered to be *un pijo*, a 'posh' migrant, by many *locutorio* users.

Naeem, the *locutorio* worker, does not even check the handset pictures in the leaflet. He is happy with his worn-out, ordinary Nokia, which seems to index some sort of a working-class identity. If he had a cooler one, clients would accuse him of earning too much from 'their' *locutorio*. In fact, his mobile is regularly checked out by his clients, and whenever he buys himself something new, a T-shirt, for instance, he is strongly reprimanded for being wealthier than them. Rachid, the *locutorio* trainee from Pakistan, by contrast, points to a phone which is specially designed for polyphonic sound management and for music storage, and which includes big headphones and a large screen, presented, in the leaflet, by young men wearing baggy trousers and DJ-ing.

Mobile phone ringtones can also serve the purposes of a staged competition for a place in the *locutorio*'s public soundscape, for example between a Spanish- and an Urdu-defender. When I ask Jenny and Shabbir what music they like, they both pick up their mobiles and start playing their favourite songs, shouting their names, almost elbowing each other out of the way. Jenny chooses the Andalusian David Bisbal (*Bulería* or fast flamenco), perhaps a symbol of modern

Spanishness, and Shabbir waves a pop song by Roxen ('Hope', in Urdu) which invokes modern Pakistaniness, while nearby computer users complain about this improvised cacophonic dispute.

The same happens with screensavers. While Sheema sends me a religious MMS with a picture of the Earth surrounded by a series of Arabic letters that read 'Allah', Naeem downloads a picture of pop singer Enrique Iglesias, which serves as one of the many postmodern digital identities that he aligns with and that he presents via his mobile, and via the *locutorio*'s computer screensaver.

Other ICT devices are also appropriated and mobilised for identity enactment purposes. James, the protest poet from Cuba, for instance, always despises the *locutorio*'s rudimentary technological infrastructure because, with its very basic software, he cannot manipulate .giv files or upload some of his work online. Besides, most USB ports are non-functional, and headphones are scarce and do not always function properly here. '¡Sólo tiene programillas para niños!' ('It only has silly little programs for kids!'), he complains, in turn positioning himself as a schooled, better-educated and more modern technoliterate web-surfer.

Likewise, the time and the money spent on the internet is also a meta-discursive terrain on which to mobilise class-related identity attributes among migrants, and where to issue some moral judgements on particular individuals' transnational trajectories. On certain occasions, for example, the *locutorio* users who employ computer services are criticised either for being 'better-off' or for being time and money 'wasters' indulging in the vices of the host societies, instead of winning their transnational bread. Yousaf, the electrician born in Pakistan who used the *locutorio* only to purchase top-ups and international landline cards emphatically sets out this particular point of view in Excerpt 6.2.

Excerpt 6.2. The construction of internet users as immoral time and money 'wasters'

@Location: 22 August 2008. Bar near a *locutorio* in El Paso. Vallès Occidental.
@Bck: When the researcher (RES) asks Yousaf (YOU) about the use of the internet, he categorises frequent web-surfers as time and money 'wasters'.

01 *RES: y a usted le gusta internet¿
    %tra: and do you like the internet¿
02 *YOU: no.
03 *RES: no¿
→ 04 *YOU: este <este de> [//] esta cosa de que no tiene trabajo que no quiere trabajar cuando [él/ella] juega con internet.
    %tra: this <this of> [//] this is a thing of those who don't have a job who don't want to work when [s/he] plays with the internet.
05 *RES: vale.
    %tra: ok.
→ 06 *YOU: este tontería pero no me gusta tontería.
    %tra: this foolishness but I don't like foolishness.
07 *RES: <no> [¿] <y> [/] y: tu <fam> [>] [¿].
    %tra: <no> [¿] <and> [/] a:nd your <fam> [>] [¿].
→ 08 *YOU: <ni un minutos> [<].
    %tra: <not a minutes> [<].
    [...]
09 *RES: no¿
→ 10 *YOU: mejor que dinero tirar.
    %tra: better to throw money away.

For Yousaf, the internet is something for people who do not want to work (line 4) and who aimlessly play and waste time instead. It is 'foolishness', which he completely disapproves of (in line 6). He claims that he would not surf the net, not even for a single minute or for family contact (line 8), and he very assuredly states that it is 'better to throw money away' (line 10). Because he is given the floor to become the storying self in an interview that was conducted on the premises of an empowering migrant-regulated institution, he constructs an argument by means of which he positions himself as a 'proper', morally righteous migrant who, unlike those who work in the underground economy and/or who do not send remittances home, is fulfilling his family duties as the head of a transnational family, and as a Muslim: 'And if you don't send money, what did you come here for¿' he later concluded in our interview.

Fighting Linguistic Exploitation: The Language and Identity Resources of the Abused

In the previous section, I have shown that inhabiting and sharing a migrant-governed alternative space of survival, resistance and subversion also means being ready to endure complex relationships based on 'conflictive togetherness', the worst consequences of which are exploitation and exclusion from one's own social networks, as I argue in this last part of the chapter, where I focus on Naeem's reactive workplace roles and myriad communicative strategies inside the discursive space of the *locutorio*.

I argue that Naeem's case illustrates how the degrading social categories attributed to language workers are triggered by a demanding, powerful mass of clients who try to access *locutorio* resources by exerting pressure upon, and even attacking, these overworked social players, which is something that such clients feel unable to do within similar venues like the call shops run by the multinational ICT companies, perhaps because of the greater likelihood of surveillance, for instance in the form of CCTV. I also show that, as a defensive and protective reaction to this, *locutorio* employees mobilise, for their survival, two main identity facets: on the one hand, they make use of a professional, serious and dry institutional character at the workplace, which is frequently played out through unmoderated language tones; and on the other, they appeal to their pauperised, abused 'have-not' persona, which is enacted with a soft voice or an apologetic tone – mostly in Spanish.

Firstly, I analyse the regime of clienthood in the *locutorio* of El Paso in detail. I show how clients are ICT-seekers who actively voice and direct their fights against marginalisation to extremely vulnerable migrant language workers, who are forced to staff the *locutorio* without any social recognition or legal protection. I then claim that the tremendous work pressure that this entails for the staff is a direct consequence of the capitalist dynamics of the globalised new economy, which allows for the institutionalisation of a new form of linguistic exploitation, today exerted among migrant entrepreneurs. This exploitation is based on a way of doing business which gives migrant employers the power to engage in very flexible, aggressive market rationalities that silently skirt many labour rights and regulations, and which allow them to exploit, to the full, their employees' unique multilingual capitals and shop assistant identity resources, as Naeem's story exemplifies.

Naeem has the profile of an average *locutorio* worker in the metropolitan area of Barcelona. About two years before I met him, he was hired by a documented Pakistani venture capitalist in his twenties with whom he had no kinship or friendship ties. Naeem worked every day of the week for more than 12 hours, with no holiday breaks, no days off and no sick leave.

However, his superior, who also owned three other 'ethnic' businesses run under three different registration names in two other towns, registered him, as an employee, with the social security administration as only a part-time worker, in order to minimise his tax bill. Naeem was paid no more than €800 a month, on a random day, at times in small instalments (none of the other *locutorio* workers whom I met in Vallès Occidental stated that they had been paid more than that). He was frequently paid late, and often less than the stipulated gross income in compensation for any 'missing' cabin call or international call card payments.

Apart from assisting customers, Naeem, like most of the *locutorio* workers whom I observed in El Paso, was responsible for getting the computers and programs booted up and functioning properly first thing in the morning. He was also in charge of: calling his boss when the account was about to run out of credit for top-ups; haggling with the call cards distributor; and calling the money transfer agency if there were problems. Besides, he had to maintain a 'proper' public shop front, which included removing the litter on the desks, cleaning the toilet, emptying the bins with the receipts and sweeping up the cigarette butts.

Naeem could never leave the front desk with the cash register, and so he always had to eat quickly, standing there on his own. He could use the toilet only when the boss or his deputies or one of his very few friends volunteered to man it (he gave them cigarettes in compensation). Normally, Naeem's superior visited (or sent a compatriot to visit) the *locutorio* only once a day, at night, in order to supervise the cashing up, to collect the total take and to give Naeem instructions for the following day. If there were problems with the accounts, Naeem was not allowed to leave until everything was sorted out. Therefore, it was not uncommon for him to stay at the workplace until well after midnight (1:30 am was the latest time I registered).

Naeem held his temporary residence permit but still needed two more years of *proved* work to get his permanent residence visa. This may partly explain why, during his time at that particular *locutorio*, he endured his superior's abuse and did not seek the free legal advice services offered specifically to migrant workers by (town-hall dependent) trade unions, a service that was available to him at a five-minute walk away. (Despite the fact that a Catalan Association of *Locutorios*, with about 2000 members, was created after the time of the fieldwork, in May 2013, the *locutorio* sector in Spain is not unionised – see *Ara*, 2013; Confederació de Comerç de Catalunya, 2013; *El País*, 2013.) The owner took advantage of Naeem's citizenship status, and of his fear of being readily replaced by other blue-collar migrants, to establish a crude top-down work regimentation scheme based on cost-efficient, profit-maximising, advanced liberal micro-business arrangements.

He could manage and supervise the worker (and the business as a whole) at a distance; he could in turn minimise costs with a single cheap, essentially trapped employee; and he did not have to deal with clients or to conduct any of the exhausting language work. Moreover, he seemed to be oblivious to the efforts required to adapt to the local sociolinguistic regimes there. On paper, his business was legal: the commodities and consumer goods that he sold were all licit, the *locutorio* licence was in order, the numbers registered on the computer were all clear, and he held proof of Naeem's registration with the social security department.

This common managerial scheme for *locutorios*, though, would not work without the exploitation of the workers' unique non-elite multilingual capitals or without their vast communicative repertoires, which are far more important than any computer knowledge or technology expertise (this is the reason why I call this sort of exploitation *linguistic exploitation*), as I now illustrate.

In the particular *locutorio* of El Paso, the owner decided to hire Rachid, a highly skilled Pakistani technician who, after a period of time as a *locutorio* trainee, was supposed to substitute for Naeem, who had never received any formal training on the functioning of the new communication technologies and who, in fact, was not interested in the digital world. In a few days, Rachid installed new programs, debugged several computers and discovered new options for the fax machine. And yet, he did not manage to operate within the particular market of the *locutorio*, basically because clients did not give him the time to develop and to work on his knowledge of the local normativity regime in the 'everyone's Spanish' translinguistic lingua franca of the place. Although Naeem kept excusing him for being 'rude', 'slow', 'inattentive' and 'uninterested', the migrant clientele put so much pressure upon Rachid that he was gone in less than three weeks.

By contrast, Naeem learnt, soon after his arrival in the neighbourhood, that his job was basically about language, and that his role was about navigating and surviving within the migrants' self-established sociolinguistic hierarchies, on their own terrain. His ability to adapt to his clients' language rules and his capacity to conduct the *locutorio*'s language work successfully was, simply, impressive. He made every effort to adjust to, and to handle, the clients' demands and complaints very quickly and efficiently, and he also learnt to read silences, to hide non-comprehension, and to foresee or to infer meaning from all sorts of misunderstandings and communication breakdowns. And he developed a set of communication weapons and identity armours to strategise with his attributed subject positions, being in turn sensitive to, and aware of, the complex power dynamics established among diverse competing or rival migrant groups of clients.

Table 6.1 Constructions of identity in discourse: Terms of address for a *locutorio* worker

Terms of address	Description and usage
Naeem	Cordial routinised term of address used by Pakistanis with whom Naeem holds a productive client–worker relationship, and by clients who have become friends
Paki	Abbreviation of *Pakistaní*, employed by young acquaintances of non-Pakistani origin (not friends), normally male and female locals, Latin Americans and Moroccans. Pejoratively used as an 'othering' label with ethnic connotations, too
Amigo Tío	'Friend' or 'mate'/'man', common terms transmitting familiarity, articulated by non-Pakistani male and female adults, and by those who find it difficult to remember Naeem's name
Jefe	Equivalent to 'boss', employed by Spanish-dominant clients acknowledging Naeem's institutional role as a *locutorio* manager. Alternatively mobilised to mockingly elevate his occupational rank by clients seeking to subvert the *locutorio* rules
Nae	Spanish-sounding abbreviation of Naeem, which transmits intimacy and closeness, employed by young migrant and local friends whom he trusts
Primo Artista	Equivalent to 'brother'/'buddy' (the second term literally meaning 'artist'), infrequently used by Spanish-dominant men seeking to assuage a demand for help
Boludo Tonto Feo	Pejorative adjectives meaning 'jerk', 'dumb' and 'ugly', used mostly by locals and Latin Americans to put the blame for a communicative breakdown or a technological problem on Naeem's institutional self
Pedazo de maricón Nenaza	Gendered insults meaning 'faggot' or 'sissy' employed by Spanish-speaking clients and by Moroccan men to ridicule Naeem's persona and to dispossess him of heteronormatively conceived manly capital
Moro de Al Qaeda	Offence with ethnic and religious connotations equivalent to 'Moroccan Al Qaeda terrorist', very infrequently articulated by local adults and non-Pakistani adolescents

Source: Field notes. *Locutorio* of El Paso, 2007–09

A useful way to reveal the pressure that these imperious *locutorio* users exert upon workers is to analyse the manner in which this clientele discursively constructs such marginalising employee identities in public, via carefully chosen terms of address. In Table 6.1 I provide the different social categorisations and the workplace roles ascribed to Naeem, articulated by the *locutorio* users of El Paso (an average of 61 to 156 clients per day), organised by frequency of use, in descending order.

Table 6.1 shows that a *locutorio* employee, if not constructed as a mere nameless worker, may be called at least 10 different Spanish nicknames. He may emerge as a respected, legitimate member of a heterogeneous network of youngsters, and even as a close friend for some of his compatriots. In turn, he may become the centre and the source of mockery, particularly when ethnolinguistically distant clients (mainly locals, Latin Americans and Moroccans, in Naeem's case) try to negotiate access to the *locutorio*'s transnational wealth (material things like cheaper bills, spare coins and cigarettes, and also symbolic resources, like legitimacy). Clients tend to attack the worker by taking the competition into the domains of physical appearance (with attributes like ugliness, as in *feo*) and of intelligence and cognitive abilities (like dumbness, as in *tonto*), on the one hand, and into the terrain of gender and sexuality (by mobilising pejorative homosexuality attributes, as in *maricón*) or of terrorism (as in *moro de Al Qaeda*), on the other.

As a step further, in Snapshot 6.3 (which gathers field notes from 9 July, 10 August, and 1 and 4 September 2008), I present some of the difficult encounters that Naeem had to deal with in his alternative front-line customer services, in order to investigate more closely the ways in which migrant *locutorio* workers manage clienthood regimes and react and fight against the attributed social roles emerging from the terms of address that I have just described above.

Snapshot 6.3. Dealing with the established regimes of an empowered clienthood

On a day when a series of unfortunate events occurred one after another, I learnt that the worst thing that can happen to a *locutorio* worker is to find that, suddenly, the telecommunications infrastructure is not working. Naeem can climb onto a chair to reach the button providing the line and deactivate and activate the system again and again; it will not work until all the computers are switched off, the telephones hung up and the three programs in the main computer rebooted. All sorts of complaints ensure then. 'Oye ¿Qué pasa con África hoy?' ('Hey what's the matter with Africa today?'), shouts a friend of the mechanic who dropped by the other night. '¡Tu, se me está cerrando!' ('You, it's switching off!'), yells a Moroccan girl from computer number two. 'Disculpa ¡Ni siquiera empecé a hablar y ya empieza a facturar!' ('Excuse me, I haven't even started talking and the counter is already running!'), complains the Ecuadorian neighbour from cabin number eight when the system starts to work again. Simultaneously, a client waiting for a fax to be sent gets nervous and calls to him with a '¡Va tío!' ('Come on, man!'), to the amusement of Luis, who keeps pestering Naeem by asking me '¿Quieres un novio feo pa que no te lo quiten? Yo te doy este' ('Want an ugly boyfriend that no one will want to take away from you? I'm giving you this one'). When adolescents come in, Naeem gets even more anxious. In the afternoon, for instance, the two teenagers who call Cuba on a daily basis frequently start banging on the glass door, throwing food all over the place and getting increasingly excited at Naeem's annoyed face. He then has to apologise to the other clients for that improvised party: 'Son chicas, yo no puedo gritar, no puedo tocar' ('They are girls, I can't shout; I can't touch'). Pressure is even worse when he has to explicitly tell clients that they have crossed the line, for some can threaten him very seriously, with expressions like '¡A mi no me vayas de chulo que te voy a romper la cabeza!' ('Don't dare bully me, I'll smash your head in!), today employed by a client who did not want to pay the entire bill. At times he has sacrificed personal resources (like cigarettes, bottles of coke and spare change), even with self-denigration, by allowing clients to collectively search inside his pockets to convince them that he is indeed not wealthier than them. The worst day, though, was when he had to chase a customer who had not paid his bill down the street (without success), for that publicised his vulnerability in the streets of El Paso, in front of this multiplicity of clients, at peak hours.

Locutorio workers who navigate these client regimes have developed a series of both protective and reactive tactics (communicative and otherwise) to keep the business running. One of the things that Naeem does in a systematic manner, for example, is to keep an eye on his monitor in order to check which cabins are engaged (marked in blue on the screen), which are not (in green) and which are occupied but idle (also in green). He issues the receipts as soon as the computer program turns the cabin numbers into green, and he places them on the desk, following the order of the row of cabins. In this way, each client's face is associated with a cabin number and its corresponding receipt. Simultaneously, he also keeps an eye on the computer desks, except for the two in the corner out of sight, which have no webcams to be monitored.

Another service that Naeem has to monitor is internet access, which works as follows. Habitual clients tend to prepay for connection time and then, when they have five minutes left, a note in Spanish pops up on the screen, warning them about the time remaining. If they want to continue using the computer, they normally have to stand up and in a hurry place some more coins on the counter, always making it clear that that particular computer desk is occupied, if there are other customers waiting for a terminal. Another modality is to shout '¡*tiempo libre!*' ('free time', in Spanish, if they are not Urdu or Panjabi speakers), which means that the user can go online for as much time as she or he needs, without interruption, and pay at the end. Naeem immediately reacts to this by clicking on the computer number and changing the client's connection status, in a second computer program, with the extra pressure that if he does not do this in a few minutes, complaints will ensue.

Naeem also attends the clients who are waiting at the front desk for a photocopy, a receipt or the like. They do not necessarily queue, for sometimes they wait in a semi-circle around him, tapping coins on the desk, or looking at him quietly but, more often than not, impatiently. Simultaneously, he also double-checks payments, assigns new cabins and computers, answers the *locutorio* phone, checks PIN numbers, provides change and keeps the paperwork organised. These activities are always interrupted by small things: a client grabbing a pen or accessing a cigarette hidden under the counter, children playing here and there, and the like.

This sort of multitasking is highly structured and organised. The pressing of the keys on the keyboard is carried out in a disembodied manner, with instinctive finger movements, without even looking at the computer. The pulling out of the receipt from the little machine next to the keyboard is almost immediate, too, as if Naeem's hands were tailored to those repetitive movements. He even remembers the telephone numbers of the most frequent *locutorio* users, for whom top-ups take just a few seconds.

Some of these activities are accompanied by one or two words in Spanish (if they are not directed to an Urdu- or a Panjabi-speaking user), repeated many times a day. These include: '¿Número?' ('Number?', for top-ups), '¿Cuántas?' ('How many?', for photocopies) and '¿Nombre?' ('Name?', for money transfers). Short greetings, farewells and monosyllabic answers abound (notably *sí*, *no* or *vale* and *bueno* for 'okay'). By contrast, both phatic talk and volubility are a marked linguistic behaviour, to be kept to the minimum. While the politeness markers 'por favor' ('please') and 'gracias' ('thanks') are frequently used, the usual shop assistant rhetoric to advise, persuade or convince customers is definitely not common here either. Instead, what language workers in *locutorios* have to command at the desk is the maintenance of a high voice volume and a convincing tone that conveys that no exceptions will be allowed concerning the breaking of the *locutorio* rules. A huge effort is required, in the form of frequent repetitions of short chunks of language (normally, instructions), described below.

'Cuando contesta pulsa almohadilla' ('Press the hash key when they answer'). This is what the *locutorio* worker tells customers when they enter a cabin, hear the call receiver's voice, but do not know what to press in order to be able to answer. By using this kind of 'safe talk' (Chick, 1996: 30), Naeem has no need to fully understand the clients' complaints or questions, for he already knows that the hash key is a source of problems, and so he effectively anticipates the solution with this sentence, displaying engagement in the service encounter process without necessarily following his clients' words. The chunk 'cuando contesta pulsa almohadilla' has been heard so many times inside the *locutorio* that even the clients chorus and employ it themselves in order to help new users.

An attenuating rhetoric (Boutet, 2006: 77) with flattering adjectives like '¡Guapa!' ('Beautiful!') for females, or polite modes of address such as 'señor' ('sir') for males, uttered with a soft, shy voice, is what Naeem uses when he needs to invoke an apologetic tone in order to stop face-threatening language or to skirt the imminent breakdown of a hitherto smoothly running afternoon (again, he uses this to non-Pakistani customers). Emphatically paired imperatives (instead of reformulations or paraphrases) like '¡Cambia! ¡Cambia!' ('Change! Change!') or '¡Espera! ¡Espera!' ('Wait! Wait!') are employed as conversation fillers to forestall non-individual complaints and to provide some unspecified explanation for a problem while this is actually still being resolved.

'No se puede' ('It's not allowed') or a more personalised 'Amigo yo no puede' ('My friend I [am] not allowed') he repeatedly uses when clients try to break the rules, for example when they ask him to hide a stolen mobile under the counter. Although this normally works, he can get very aggressive responses back when he uses these two expressions.

Another tactic Naeem uses in monitoring the shop is to appeal to his agentless and non-specialist, non-institutional self by presenting himself as a mere employee trying to establish some common ground among 'equals'. No day goes by without clients asking him to change bank notes. Since he needs many coins to deal with small purchases (particularly for top-ups), he has developed two linguistic resources to avoid running out of change. Naeem either gives users a candid explanation, with a 'Tengo pero yo quiero cambio' ('I have [some] but I want change'), or else humbly begs for understanding in a very apologetic tone and a sad face: 'No ... Perdona' ('No ... I'm sorry').

As a smart language worker, he has also established his defensive resistance system, which consists of either a firm 'No' or a 'No hay' ('There isn't any'), or a resolute 'No sé' ('I don't know'), employed to put an end to a conversation without giving customers time to retake the floor. For example, after having spent an hour informing his clients that they will have to wait for another hour until one of his workmates puts credit into their top-up balance at the bank nearby, he finishes the day with short, blunt negatives without further explanation: 'No hay recarga' ('There are no top-ups'). He uses the same type of language but as a lie when he does not trust a client, for example when somebody approaches him with a €50 note (if his boss finds that counterfeit money has been received, the sum is docked from Naeem's pay).

Naeem's most useful resource in dealing with misunderstandings and in providing counter-arguments to clients who try to cheat him is the main computer, whose monitor is strategically placed on the desk yet sufficiently visible for clients to check information on it, or for Naeem to point to the price, number or cabin without the need to use any language. (As in many *locutorios*, the mouse and the keyboard are always placed under the counter, out of the reach of clients.) In addition, a *locutorio* worker needs to make recurrent use of four instruments that help to certify, and provide evidence for, the expenses incurred by customers. These four tools all use written language in Spanish: (1) the individual call receipts, (2) the fax confirmation messages, (3) the telephone counter equipment placed inside the cabins telling users exactly how many seconds they have used up, and (4) the computer program which records internet usage and which registers all the information concerning the calls (see Figure 6.2).

In spite of the range of linguistic resources and communicative tactics, and in spite of the helpful technological infrastructure that he could make use of, Naeem was the easy target of the *locutorio* users' frustration, and he even experienced overt racism and physical aggression. In the end, as a consequence of the large amount of language work that he had to do in a place where the owner monitored his salary (and his visa!) and where clients had the last word, he ended up suffering from lack of autonomy, social isolation, eating disorders, compulsive smoking, chronic fatigue, emotional distress and frequent anxiety attacks. This was so until 2009, when he moved to another *locutorio* shop to work for a similar number of hours under similar conditions, though without undergoing non-payments.

I understand the intricate 'internal' identity fights and the complex social structuration dynamics that I have presented in this chapter – analysed through the unique lens of discursively tokenised social categories such as the Moroccan 'thieves', the *Latina* 'scroungers', the 'un-Muslim' Pakistanis

Locutorios as Migrant Spaces of 'Mismeeting'   169

(a)

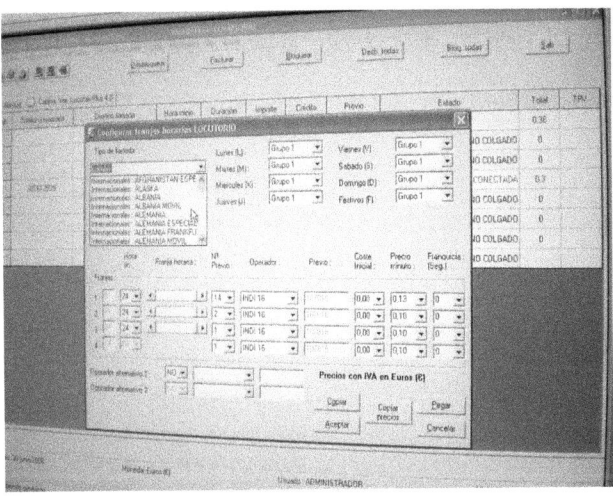
(b)

Figure 6.2 Technological support for *locutorio* employees. (a) Cabin telephone counter and (b) *locutorio* computer programs. *Locutorio* in El Paso. Pictures taken by author, 30 June and 29 July 2008

and the like – as a reflection of the dirtiest facet of alternative institutions of migration and transnational living, epitomised, above all, in the figure of an alarmingly unprotected *locutorio* worker, Naeem.

We now have evidence that linguistic exploitation is almost inherent to the crude capitalist machinery of the globalised new economy, and we also know that, of course, this sort of abuse founded on the employees' communicative capitals is by no means new. And yet, by analysing the opprobrious experiences that the language workers of El Paso endured, I have become increasingly aware that blatant exploitation in these alternative workplaces may be gaining momentum and may perhaps be starting to become institutionalised in the context of the Catalan network society. *Locutorios* may be opening a door to their voiceless army of multilingual mediators, the cheapest labour force of late capitalism, becoming a stagnated infra-class of migrants ultimately confined to the realms of 21st-century ostracism.

By Way of Conclusion: Informal Migrant Shelters in Which to Critically Explore the Mundane Alphabets of the Future

> T'escoltava a través de tertúlies d'alè de vinagre
> on mentir és un verb conjugat en honor als diners;
> per a tu construeixen barreres i murs franquejables
> per la fam, el dolor i la por que t'empeny a seguir.
> Et promet una terra fecunda d'olor a ginesta
> on fundar el demà i inventar alfabets de futur.
>
> I heard about you in those gatherings which smell of vinegar
> where lying becomes a verb to be inflected for the sake of money;
> for you, they build barriers and walls which are surmountable
> by the hunger, the pain and the fear that force you to keep going.
> I promise you a fertile land with the scent of broom
> where to establish a tomorrow and to invent alphabets of the future.
>
> Feliu Ventura, 2006 (from the song 'El pes d'un somriure' from *Alfabets de Futur*)

Locutorios epitomise both the sweetest and the bitterest fruits that the convoluted dynamics of the globalised new economy have borne for the transnational migrants of the 21st century. They are the product and the process of the internationalised market-driven governmental forces which build fences at borders in order to keep undocumented populations entangled and placeless, but at a time when these particular mobile social groups simultaneously have the potential to successfully challenge, resist and contest nation-state power in an unprecedented manner.

Locutorios also speak of the equally subversive strategies that migrant networks deploy to defend themselves against, and, as clever capitalist selves, to take advantage of, the aggressive market rationalities of the telecommunications sector. Besides, they also reveal the sophisticated ways in which they oppose the dominance and hegemony of private (and privatised) multinational capital, which, after decades of market underestimation, is now reinventing its commercial strategems in order to target them as the most lucrative consumerist market segment of the commmunication age.

In this sense, *locutorios* embody new and distinct ways of understanding, organising and practising grassroots transnationalism. Different groups of migrants have started to stand up against anti-migrant policies and anti-migrant practices in their host societies. With an increased degree of social agency enhanced (but not readily propelled) by their mass access to ICT, these populations have jointly managed to appropriate the small local shops that they now run and frequent and transform them into defensive refuge institutions which provide crucial access to the networking capital – the 'networks of resistance' (Castells, 2012: 8) – that, in today's uncertain world, ensure shelter and secure transnational survival.

Of course, we are not witnessing the birth of convivial or harmonious migrant settlements that articulate the fight against outright social exclusion. Instead, we are observing the emergence of self-sufficient, resilient migrant groups – characterised by a concomitant trust/love and distrust/enmity – who, out of necessity and perhaps despair, strategically compete for the colonisation of these places side by side but not eye to eye, in an interested coordinated togetherness based on highly hierarchised 'internal' social structuration practices and social stratification processes.

We are not talking about a well orchestrated or carefully planned type of resistance, but rather about a very pragmatic series of mundane (though internationalised) local reactions against individual daily experiences of marginalisation. This struggle, as their expertise in mobilities and in moorings has proved, is better fought collectively, under the protective umbrella of a more and more visible, and more and more audible, 'macro' group.

Thus, *locutorios* are definitely not egalitarian 'anti-system' spaces of self-conscious rebellious activism, though in manifold ways they combat, eagerly, opprobrious dataveillance systems for their users, with 'micro' tactics that include, for example, the selling of SIM cards to the vulnerable undocumented. They are not public, altruistic or charitable 'asylums' at the margins working for the ideal democratisation of the shrinking welfare state either; yet they cater for the vital material and symbolic resources of delocalised or pauperised migrants. These resources include not only cheap international phone calls and informal loan services but also legal advice, job offers, rooms for rent and even bodily safety and sustenance, like food and water. Neither do they constitute a classic entrepreneurial branch of the telecommunications sector, although their owners, who were the first social actors to understand migration as a key asset to be exploited, certainly make good money via the selling of migrant-tailored information and communication technology, undeniably boosting and contributing to the local (grey) economy, with purely capitalist, top-down managerial business structures. In short, *locutorios* materialise the emergence of the type of *hybrid alternative institutions* of transnationalism which, emanating from the periphery, and being neither totally public nor completely private, have become altogether different, fully fledged *pivotal relocating centres* for migrant populations who vindicate their local place in the global map among themselves, in connection with, but outside, hostile institutional realms.

In the Catalan context, the *locutorios'* quick consolidation has been extraordinary, to the extent that, seemingly overnight, they have exceeded all expectations, overwhelming local authorities with their geographic, economic and social expansion. This, at times, has brought about overt conflicts with local town halls, which have not been ready to deal with 'foreign' populations who, while following the established bureaucratic machinery, have started to question the rules of the game by *migranticising* (i.e. by transforming and appropriating) the landscapes of urban towns and global cities. Their prosperity has been such that, in a decade or so, *locutorios* have become the migrants' favoured place at which to access communication technology, consequently propelling, too, the gradual *informalisation* of neighbourhood economies, even overshadowing multinational enterprises like the giant Telefónica, at least with regard to international voice telephony and remittance services.

I attribute this rapid consolidation to the singular conditions in which *locutorios* were set up. Their success cannot be understood without taking into account the fact that they first emerged when most migrants experienced a fast 'technologisation' of their transnational lives and prospects (particularly through the use of mobile phones in combination with prepaid SIM cards), which allowed for the self-management of their access to the global telecommunications grid, in Spanish territory. This technological mobilisation turned into an unparalleled social revolution, which spread, in a rhizomatic way (i.e. en masse, horizontally and vertically), within and across migrant social networks of almost all socioeconomic, cultural and linguistic backgrounds.

It was under these circumstances that new, documented, businessminded migrant entrepreneurs, who understood that their ventures had to gather, handily, all sorts of social media devices within one single locale, operating below the radar of the registration authorities, turned *locutorios* into the migrants' cradle of 'communication as resistance' (Horst & Miller, 2006: 15), that is, into spaces in which to defiantly collectivise and share ICT gadgets and to redistribute networking capital across transnational family units, at the margins of the instituted circuits of ICT commerce.

The second reason for the *locutorios'* commercial success concerns, crucially, language. The most discriminating governmentality constraints against which migrants struggled and struggle today are constructed around an old reactionary linguistic regime which, in new European 'civic' guises, and with neoliberal globalist discourses seemingly acknowledging linguistic diversity, keeps investing in monolingualism and keeps establishing a social organisation of life in and through single nation-state languages. Indeed, such monolingualism has become a criterion for the granting of full citizenship rights.

In Catalonia, this means that, at present, the machinery of the techno-political governmentality block constituted by the Spanish government and by the telecommunications market is geared towards the imposition of a linguistic uniformity in Spanish. From this perspective, dominant European languages (particularly English, unproblematically welcomed as the pioneering code for conducting global business and for borderless 'intercultural communication') have a relatively small presence; co-official languages like Catalan are in general publicly made non-public, despite genuine attempts on the part of the Catalan government to protect it and to sell it in modern economicist terms; and non-elite (i.e. devalued, silenced) allochthonous minority codes, from Tamazight to Panjabi, but also from Arabic to Urdu, are, simply put, almost non-existent.

Perhaps in an unintended manner, *locutorios* have challenged these language frames based on nation-building projects and have ended up redressing and combating this prejudiced, discriminatory linguistic landscape. Thus, one rationale behind *locutorios* today is the functional, rudimentary connection of the many migrant populations who, being non-schooled in or at least not familiarised with the prevailing numeracy and literacy standards of the north, emerge as the new technology 'have-nots', the global illiterates of the information era.

The real articulators of these alternative institutions are, undoubtedly, the overworked *locutorio* employees who, with their routinised unpaid language work, provide the complex technoliteracy capitals and the informal sociocultural mediation, transliteration and translation which are necessary for migrants to insert themselves into the bureaucratic cultures of their host societies. And it seems that *locutorios* are here to stay, because they, at zero cost, keep doing and keep finishing the multilingual job which the nation-state, the telecommunications sector and the voluntary sector have left not only unaccomplished but also unattended in Catalonia.

This is what makes them a privileged venue for the diversity of the so-called 'new migratory era', which has already taken up a considerable research space within the sociolinguistics of globalisation. On the one hand, *locutorios* open up a new arena in which to examine migrants' nuanced identity work (i.e. their largely unknown intra-/inter-group presentations of the selves and of 'others', based on fluid class, ethnicity, religious, gender and age traits), which, indeed, reveals how they orient to, inhabit, enact, combat and, in short, experience transnational survival and transnational living. In this sense, *locutorios* become, more generally, a venue for the exploration, from the inside, of what migration actually means for diverse migrant populations, thus allowing us to distance ourselves from, and to problematise, non-socially engaged, pre-established etic canons concerning how migrant individuals understand who they believe they are and what part of the globe they feel they inhabit.

In addition, *locutorios* allow for the study of the multi-valued and ever-changing linguistic features, social uses and social meanings of migrant voices, which consist of hybrid, truncated non-elite multilingualisms and of technology-mediated vernacular literacies and numeracies. They allow us to critically address the too frequently forgotten migrants' polyphonies of everyday life in their full complexity and with all their social contradictions and ambivalences. And they immerse all of us interested in language into situated communicative events which may speak of both challenging counter-hegemonic repertoires and of self-imposed or self-disciplining written and spoken practices into the dominant standard norms.

In short, *locutorios* reveal the intricacies of the under-researched 'internal' power dynamics established, negotiated and fought in and through these fluid relocalised linguae francae which embody the renegotiation of new affinities, new rivalries and new belongings in urban Catalonia. They provide a unique picture of the real mundane translinguistic alphabets with which social difference and, more broadly, social inequality are simultaneously actively resisted and, paradoxically, produced and reproduced by migrant global trotters who, against all sorts of adversities, have managed to colonise these non-mainstream worlds of late capitalism.

Notes

(1) I follow Bourdieu's notions of both social capital and resources (Bourdieu, 1977, 1986, 1990, 1991).
(2) I use 'Spanish state' (instead of 'Spain') to highlight the fact that, under this political territorial unit there is a heterogeneity of autonomous regions (like Catalonia, Galicia and the Basque Country) which have different degrees of self-government and which are bi-/multilingual.
(3) Following Kroskrity (2000: 3), I employ the term *regime* to stress the political-economic activities which shape those linguistic practices that characterise present-day modes of institutional top-down governance and citizenship control.
(4) I use the term 'host societies' (rather than 'host society') to refer to the heterogeneous Catalan and/or Spanish-speaking people who, born anywhere in Spain, live in Catalonia.
(5) I understand social agency to be the capacity to react to and to act upon exclusionary institutional regimes, following Giddens' (1984) and Moyer's (2011) frameworks.
(6) I use the term 'structuration process' in Giddens' sense (Giddens, 1984; see also Heller, 2007b).
(7) By linguistic ideologies I mean the multiple, frequently contradictory or ambivalent indices of the norms, judgements, positionings and interests governing the migrants' given sociolinguistic behaviours which play a crucial role in the construction of meaning (Blommaert, 1999; Heller, 2007c; Irvine & Gal, 2000; Schieffelin *et al.*, 1998).
(8) In this book, I use sometimes the terms *multilingualims* and *non-literacies* in the plural, in alliance with Martin-Jones and Jones' (2000: 4) critique of 'the a-social, a-historical skill/ability understanding of reading and writing' which excludes or renders invisible non-Western-regulated ways of practising language, literacy and numeracy.
(9) See similar uses of this term in Alarcón and Garzón (2011a), Glick Schiller and Çağlar (2013), Panagakos and Horst (2006), Piller (2011) and Ros *et al.* (2007), by way of example.
(10) The terms 'ethnic business' and 'ethnic entrepreneurship' were coined by Ivan Light (1972), and were then taken up by other pioneering American anthropologists and sociologists like Roger Waldinger (Waldinger *et al.*, 1990). In the mid-1980s, scholars from fields as diverse as geography, economics, business management, anthropology and sociology jointly adopted and systematised the use of these terms to analyse

the expansion of (low-class, minority) 'immigrant' businesses in Europe and North America, where the migrants' entrepreneurial initiatives had then started to gain salience and visibility. Since then, research on 'ethnic commerce' has developed into a discipline of its own and has become a central issue of analysis in both migration and economic studies – see the collection in Dana (2007), by way of example.

(11) All quotations, bureaucratic documents, advertisements, newspaper quotes, songs and visual materials in Catalan and Spanish have been translated by the author. The materials in languages other than Catalan or Spanish have been translated by Marc Carreras, Najma Husain, Youssef Ihmadi, Safae Jabri, Nargus Karim, Daniel Ramon, Pimpila Thanaporn and Lin Xie. All quotes in their original languages are provided in the endnotes. Original quote here: 'Aquellos comercios regentados por personas de origen extranjero [...] con independencia de las características de los servicios o productos que se mercantilizan y de la distribución geográfica de estos establecimientos.'

(12) See the normalised use of the terms *negoci ètnic* – ethnic business, in Catalan – and *empresariado étnico* – ethnic entrepreneurship, in Spanish – by Beltrán *et al.* (2007), Cebrián de Miguel and Bodega Fernández (2002), Parella Rubio (2005), Serra del Pozo *et al.* (2003) and Solé *et al.* (2007).

(13) This representative sample included a total of 2,481,444 households (of different socioeconomic backgrounds), with 484,563 people of foreign origin and 4,970,869 Spanish nationals living in Catalonia.

(14) Directive 2008/115/EC of the European Parliament and of the Council of 16 December 2008 on common standards and procedures in member states for returning illegally staying third-country nationals. Online at http://eur-lex.europa.eu (accessed 13 July 2013).

(15) This includes: self-employed domestic workers, self-employed farming and agriculture workers, self-employed sea workers and self-employed miners (Observatorio Permanente de la Inmigración, 2010).

(16) Technoliteracy is here defined as the command of a complex set of visual, auditory, iconic, digital, informational, technological and computer literacies required to access the global ICT system successfully (Area Moreira *et al.*, 2008; Kahn & Kellner, 2005).

(17) Exact percentages are not provided for confidentiality reasons, since the localities could easily be identified from the data.

(18) 'Identificar al dueño de cada número es una tarea de vital importancia en la lucha contra el terrorismo y el crimen organizado.'

(19) 'Evidentemente la tecnología no entiende de idiomas; es la misma para todo el mundo.'

(20) The Madrid bombings in March 2004, for which 29 people from Morocco, Syria, Algeria, India and Spain (among others) were tried, were apparently carried out by three Moroccans (one of whom, Jamal Zougam, had already been convicted for murder, theft and terrorism in 2001) and by two Spanish-nationalised Indians who activated the bombs using prepaid mobile phones.

(21) '[...] para la protección de las personas y bienes y para el mantenimiento de la seguridad pública; [...] fines delictivos.'

(22) '[...] su preferencia inicial es un teléfono de prepago porque no requiere facilitar los datos personales ni la cuenta bancaria.'

(23) The emphasis on using Spanish as the public language of the administration and as the resource to access Spanish citizenship was also attested by Codó (2008) and by

Codó and Garrido (2010) in their ethnographies of a Catalonia-based state office for migrants undergoing legalisation processes and a free legal advice service offered by a pro-migrant, non-profit organisation.

(24) By *privatised* I mean those formerly state-owned resources (normally, goods or services employed by the vast majority of the population, such as telecommunications, water, electricity, fuel or gas services) that later became market assets and thus started to be sold by individual companies, according to the rules of the market, despite the fact that in several cases they are still directly or indirectly surveyed and constrained by the government, albeit often indirectly.

(25) I argue that it is a crisis-resistant sector because, according to the latest official statistics, the business volume that this particular market generated worldwide accounted for a 12% of the world's merchandise, or US$1.5 trillion, corresponding to 2.4% of global GDP, in 2010 (ITU, 2012a: 127).

(26) According to the Spanish Ministry of Industry, Tourism, and Commerce, at that time 88% of households used Movistar's infrastructure, and 79% employed its telephone services. Besides, the multinational handled 77.2% of local and 64.8% of international calls (Yagüe Llorente, 2005).

(27) In Catalonia, the numbers for mobile phone ownership were even higher; in Barcelona city, for instance, there were 2.3 mobiles per household already in 2008 (*L'Hiperbòlic*, 2008: 12–13).

(28) 'muy superiores a los que proporciona un cliente autóctono.'

(29) '[...] aportar información de mercado objetiva y recurrente en un nicho, el de los inmigrantes, que está cobrando una importancia estratégica trascendental.'

(30) '[...] para la promoción de la *integración* [...] y para *la comunicación entre el sector empresarial y este colectivo concreto de la población*' (my emphasis).

(31) '¡Ni de autónomo ni nada! No quieren ni ni ver papeles ... quieren ver el dni' (quote from the interview reproduced verbatim).

(32) 'A cada cual en su idioma. Esa es la clave para llegar a la comunidad extranjera en España.'

(33) Language economists are economists investigating how economic and linguistic variables influence one another (Kamwangamalu, 2008).

(34) This information was provided by a professor of translation and interpretation studies and IBM researcher at the Barcelona 7th Telecommunications Day in Catalonia, 30 September 2008.

(35) These were Jazztel, XL Móvil, Blau, Hits Mobile, FonYou, Carrefour Móvil, Día Móvil, Happy Móvil, LlamaYa and BT Móvil.

(36) 'Per un immigrant aquí pues potser no li cal, no? i diu "pues per què haig de gastar-me mig euro en traduir algo," no? "ja li preguntaré a qui sigui".'

(37) 'Cuando mi jefe me explota ¿Quién me escucha?,
sin papeles no hay trabajo y sin trabajo no hay papeles,
espiral extraña y España me atrapó en sus redes.
¿Qué puedo hacer? Respiro desesperanza,
malvivo en un zulo oscuro porque mi alquiler no alcanza,
acudo al locutorio a diario, ese es mi deber,
saber cómo están los míos, qué tendrán para comer.'

(38) In the late 1990s, for instance, a call to Morocco cost the equivalent of €1–1.20 a minute with Telefónica, whereas it cost 58–72 cents with those first Spanish *locutorio* companies which were supported by powerful telecommunications partners overseas (*El País*, 1997).

Notes   179

(39) 'El fenómeno de los locutorios está ligado a la inmigración. El *boom* empezó en el año 1999 y ha explotado ahora.'
(40) Telefónica, which was losing its hegemonic control over the international telephone services market, sued these emergent *locutorio* companies for unlawful competition, to no avail. After several complex trials, Comytel proved to the CMT that the former monopoly was purposefully delaying or refusing to set up new telephone lines to the *locutorios* that did not use its infrastructure or that did not sign an exclusive rights contract with Telefónica (*El País*, 2003), for which the multinational was finally admonished. After this, the CMT closely monitored Telefónica's commercial practices.
(41) In all cases, applicants had to provide proof of legality, register in the Treasury's record of compulsory taxpayers, fill in (and pay for) a licence form, and, finally, deliver to the municipal government a technical report with a property plan certified by a collegiate architect.
(42) In the well known migrant-populated neighbourhood of the Raval in Barcelona, for instance, the number of *locutorios* increased from 55 in 2001 (Moreras, 2007: 55) to more than 100 by June 2010 (pro-migrant association IBN-Batuta ASCIB, 15 June 2010, personal communication), mostly started by Pakistanis. In Sant Pau Street alone, in this same neighbourhood, the number of *locutorios* had reached 24 by 2012 (Pagliarin, 2012).
(43) The diverse town council sections included Urban Planning, Territorial Services, Community Involvement, and Economic Activities. The different, unsystematised registration names consisted of *Locutori, Locutori Internet, Assessoria i Gestió Locutori, Centres d'Internet, Locutori i Activitat d'Informació, Serveis Telefònics per a Ús Públic* and *Cíber Locutori*.
(44) Personal communication by the town hall of Sabadell, 13 January 2009.
(45) 'Els estem posant unes condicions que són pràcticament impossibles de complir.'
(46 'L'èxit i proliferació d'aquests serveis juntament amb l'absència de legislació tècnica específica provoca un increment de molèsties i activitats paral·leles que poden concloure en una disminució de la qualitat de vida dels ciutadans.'
(47) 'Els locutoris, resultat de la iniciativa empresarial de ciutadans d'immigració recent, [...] posen en circulació informacions rellevants per a la vida diària i la incorporació a la ciutat. El locutori és més que un centre de contactes privats [...]. És un local on es poden trobar coneguts i on s'encreuen informacions personals i col·lectives.'
(48) See, for instance, Vodafone's plan *Comunidad Mi País* (Community My Country) (Vodafone, 2013) or Movistar's offer *Tarifa Habla Internacional* (Discount Plan Speak International) (Movistar, 2013).
(49) 'Tu gente a sólo un céntimo de distancia.'
(50) Original quotes (in the order in which they have been presented): 'Para llamar a casa sólo tienes que querer,' 'Un amigo es alguien con quien no te cuesta hablar,' and 'Un móvil para ellos, menos gastos para ti.'
(51) Original blog post: 'En Barcelona los locutorios [...] casi siempre son atendidos por pakistaníes que no entienden mucho de español. Parece que sólo les enseñan a decir 'una hora un euro' y luego no te dicen por favor pase a la máquina seis, sino sólo te dicen 'seis'. 'Seis qué' le dije yo el primer día y el pakistaní de turno me miró angustiado con cara de a este cuate no le entiendo ni rosca, tratando inútilmente que sus neuronas procesaran el mensaje. Tranquilo pakistaní, le dije, me voy a la máquina seis porque supongo que eso querés, pero no me pongás cara de chucho regañado pues, que no es para tanto, aquí no estamos peleando. A la siguiente vez le

dije que sólo quería media hora, que cuánto costaba. El pakistaní volvió a poner su cara de chucho regañado. Mirá pues pakistaní, le dije, si una hora un euro [...], media hora 50 céntimos, cin-cuen-ta cén-ti-mos pakistaní, ¿vale?'
(52) The Urdu script is an extension of the Persian alphabet, which in turn is an extension of the Arabic script.
(53) 'Primero el español. El español es el segundo idioma del mundo.'
(54) Other studies which attest to this fact in the urban metropolitan area of Barcelona include: Alarcón and Garzón (2011a: 141), Bastardas i Boada (2012: 77), Castells *et al.* (2004: 242), Fabà Prats (2012: 51–56), Garzón (2012: 2508), and Hernández-Carr (2011: 106).
(55) These 'dialects' are meant to be indicative of fully fledged languages. This seems to be a common naming practice of certain linguistic codes among particular groups of migrants, mainly among those born in Africa (Díaz, 2004).
(56) Alternatively, by including a final –s in the word *novides*, it may also be possible that Shabbir used the plural form for *Navidad*, which is *Navidades*.
(57) Examples of the use of the term 'immigrantese' in a censorious way are available in blog posts like Atari (2006), Axford (2009) and Topix (2009).
(58) I follow previous work on identity and social structuration processes conducted by Barth (1969), Bauman (2005), Giddens (1991), Martín-Rojo (2010), and Unamuno and Codó (2007).

References

ACPI (2007) *Estudio de medios para inmigrantes. Resumen general EMI 2007*. Online at http://www.fct.urjc.es/oicam/enlaces/docs/emi2007.pdf (accessed 30 December 2013).
ACPI (2008) *Estudio de medios para inmigrantes. Resumen general EMI 2008*. Online at http://www.fct.urjc.es/oicam/enlaces/docs/emi2008.pdf (accessed 30 December 2013).
ADN (2008) Advertisement by Hits Mobile, 16 December, p. 15 (in the Deportes section of the newspaper published in Barcelona by Página Cero S.A.).
Agar, J. (2003) *Constant Touch: A Global History of the Mobile Phone*. Cambridge: Icon Books.
Ajuntament de Rubí (2005) Ordenança municipal reguladora dels establiments de serveis telefònics per a ús public. Online at http://www.ajrubi.es/recursos/ajrubi/recursos/campanya_inspecci_locutoris_ordenansa_municipal.pdf (accessed 13 July 2013).
Ajuntament de Terrassa (2009) Ordenança municipal reguladora dels establiments de serveis telefònics per a ús public. Online at https://aoberta.terrassa.cat/normativa (accessed 13 July 2013).
Alabajos, P. (2011) El forat del pany. In Obrint Pas (eds) *Coratge* (np). Alzira: Bromera.
Alarcón, A. (2007) Informationalism, globalisation and trilingualism. An analysis of the statistics of linguistic practices in small and medium companies in Catalonia. *Noves SL: Revista de Sociolingüística*. Online at http://www6.gencat.cat/llengcat/noves/hm07tardor-hivern/docs/a_alarcon.pdf (accessed 13 July 2013).
Alarcón, A. (2011) Economia de la llengua. *Treballs de Sociolingüística Catalana* 21, 19–27. Online at http://revistes.iec.cat/index.php/TSC/article/view/53827/pdf_111 (accessed 13 July 2013).
Alarcón, A. and Garzón, L. (2011a) Concluding remarks: The instrumental and symbolic dimensions of multilingualism and comparative perspective. In A. Alarcón and L. Garzón (eds) *Language, Migration and Social Mobility in Catalonia* (pp. 139–150). Leiden: Brill.
Alarcón, A. and Garzón, L. (2011b) Language and second generation in Catalonia. In A. Alarcón and L. Garzón (eds) *Language, Migration and Social Mobility in Catalonia*. (pp. 1–13). Leiden: Brill.
Alarcón, A. and Garzón, L. (2011c) Evaluation of social mobility and language in context. In A. Alarcón and L. Garzón (eds) *Language, Migration and Social Mobility in Catalonia* (pp. 123–138). Leiden: Brill.
Alarcón, A. and Heyman, J.McC. (2013) Bilingual call centers at the US–Mexico border: Location and linguistic markers of exploitability. *Language in Society* 42 (2), 1–21.

Alonso, A. and Oiarzabal, P.J. (eds) (2010) *Diasporas in the New Media Age: Identity, Politics, and Community*. Reno, NE: University of Nevada Press.

Álvarez-Cáccamo, C. (1998) From 'switching code' to 'code switching': Towards a reconceptualisation of communicative codes. In P. Auer (ed.) *Code-Switching in Conversation: Language, Interaction and Identity* (pp. 29–51). London: Routledge.

Anderson, B. (2006 [1983]) *Imagined Communities: Reflections on the Origin and Spread of Nationalism*. London and New York: Verso.

Androutsopoulos, J. (2006a) Introduction: Sociolinguistics and computer-mediated communication. *Journal of Sociolinguistics* 10 (4), 419–438.

Androutsopoulos, J. (2006b) Multilingualism, diaspora, and the Internet: Codes and identities on German-based diaspora websites. *Journal of Sociolinguistics* 10 (4), 520–547.

Androutsopoulos, J. (2007) Language choice and code switching in German-based diasporic web forums. In B. Danet and S.C. Herring (eds) *The Multilingual Internet: Language, Culture, and Communication Online* (pp. 340–361). Oxford: Oxford University Press.

Androutsopoulos, J. (2009) Language and the three spheres of hip hop. In H.S. Alim, A. Ibrahim and A. Pennycook (eds) *Global Linguistic Flows: Hip Hop Cultures, Youth Identities, and the Politics of Language*. New York: Routledge.

Androutsopoulos, J. (2011) From variation to heteroglossia in the study of computer-mediated discourse. In C. Thurlow and K. Mroczek (eds) *Digital Discourse. Language in the New Media* (pp. 227–298). Oxford: Oxford University Press.

Anis, J. (2007) Neography: Unconventional spelling in French SMS text messages. In B. Danet and S.C. Herring (eds) *The Multilingual Internet: Language, Culture, and Communication Online* (pp. 87–115). Oxford: Oxford University Press.

Appadurai, A. (1996) *Modernity at Large: Cultural Dimensions of Globalization*. Minneapolis, MN: University of Minnesota Press.

Ara (2013) Neix el Gremi de Locutoris de Catalunya, 28 May. Online at http://www.ara.cat/economia/locutoris-suneixen-Gremi-Locutoris-Catalunya_0_927507503.html (accessed 30 May 2013).

Arango, J. (2009) Después del Gran Boom: La inmigración en la bisagra del cambio. In E. Aja, J. Arango and J. Oliver Alonso (eds) *La inmigración en tiempos de crisis, Anuario de la Inmigración en España* (pp. 52–73). Barcelona: CIDOB Edicions.

Ardèvol, E., Estalella, A., Gómez-Cruz, E. and Enguix, B. (2008) Media practices and the Internet: Some reflections through ethnography. Paper presented at the European Communication Research and Education Association Conference, 25–28 November, Barcelona. Online at http://www.slideshare.net/Estalella/towards-an-ethnography-of-new-media-practices-reflections-through-field-experience-presentation?src=embed (accessed 13 July 2013).

Area Moreira, M., Gros Salvat, B. and Marzal García-Quismondo, M.A. (2008) *Alfabetizaciones y tecnologías de la información y la comunicación*. Madrid: Síntesis.

Arjona Garrido, A. and Checa Olmos, J.C. (2006) Economía étnica: Teorías, conceptos y nuevos avances. *Revista Internacional de Sociología* 64 (45), 117–143.

Atari (2006) Atari Community Forums, blog. Online at http://ataricommunity.com/forums (accessed 10 October 2009).

Atkinson, D. (2000) Minoritisation, identity and ethnolinguistic vitality in Catalonia. *Journal of Multilingual and Multicultural Development* 21 (3), 185–197.

Auer, P. (ed.) (1998) Introduction: Bilingual conversation revisited. In P. Auer (ed.) *Code-Switching in Conversation: Language, Interaction and Identity* (pp. 1–24). London: Routledge.

Avui (2009) 800.000 catalans no han identificat encara el mòbil, 10 November. Tecnologia: 28 (*Avui* newspaper published in Barcelona by Hermes Comunicacions S.A.).
Axford, E. (2009) 'WWJD?' e.e. cumming all over your titties. Blog entry dated 22 January. Online at http://emilyaxford.wordpress.com (accessed 10 October 2009).
Aymà Aubeyzon, J.M. (2012) Els usos personals o privats a Catalunya. Balanç i perspectives de futur. *Treballs de Sociolingüística Catalana* 22, 59–72.
Barcelona Activa (2011) *Informe sectorial Telecomunicacions i TIC*. Barcelona: Ajuntament de Barcelona. Barcelona Activa.
Barnett, G.A. (2001) A longitudinal analysis of the international telecommunication network, 1978–1996. *American Behavioral Scientist* 44 (19), 1638–1655.
Baron, N.S. (2008) *Always On: Language in an Online and Mobile World*. Oxford: Oxford University Press.
Barth, F. (ed.) (1969) *Ethnic Groups and Boundaries: The Social Organisation of Cultural Difference*. Boston, MA: Little, Brown and Company.
Bastardas i Boada, A. (2012) El català i els joves: propostes de política lingüística del Consell Social de la Llengua Catalana. *Treballs de Sociolingüística Catalana* 22, 77–92.
Bauman, Z. (2000) *Liquid Modernity*. Cambridge, MA: Blackwell Publishers.
Bauman, Z. (2001) *Community: Seeking Safety in an Insecure World*. Cambridge: Polity Press.
Bauman, Z. (2005) *Identitat: Converses amb Benedetto Vecchi*. València: Publicacions de la Universitat de València.
Baynham, M. and Lobanga Masing, H. (2000) Mediators and mediation in multilingual literacy events. In M. Martin-Jones and K. Jones (eds) *Multilingual Literacies: Reading and Writing Different Worlds* (pp. 189–207). Amsterdam: John Benjamins.
BBC Mundo (2008) España: La inmigración es multimedia. BBC Mundo, 28 June. Online at http://news.bbc.co.uk/hi/spanish/business/newsid_7478000/7478177.stm (accessed 1 July 2013).
Beltrán, J., Oso, L. and Ribas, N. (eds) (2007) *Empresariado Étnico en España*. Observatorio Permanente de la Inmigración. Barcelona: Centre d'Estudis i Documentació Internacional de Barcelona. Madrid: Subdirección General de Información Administrativa y Publicaciones.
Bertot, J.C. (2003) The multiple dimensions of the digital divide: More than the technology 'haves' and 'have nots'. *Government Information Quarterly* 20, 185–191.
Blommaert, J. (ed.) (1999) *Language Ideological Debates*. Berlin: Mouton de Gruyter.
Blommaert, J. (2003) Commentary: A sociolinguistics of globalization. *Journal of Sociolinguistics* 7 (4), 607–623.
Blommaert, J. (2008) *Grassroots Literacy: Writing, Identity and Voice in Central Africa*. London: Routledge.
Blommaert, J. (2010) *The Sociolinguistics of Globalization*. Cambridge: Cambridge University Press.
Blommaert, J. (2012) *Chronicles of Complexity: Ethnography, Superdiversity, and Linguistic Landscapes*. Tilburg Papers in Cultural Studies 29. Tilburg: Tilburg University.
Blommaert, J. and Rampton, B. (2011) Language and superdiversity: A position paper. *Working Papers on Urban Language and Literacies* 70, 1–22.
Blommaert, J., Collins, J. and Slembrouck, S. (2005) Spaces of multilingualism. *Language and Communication* 25 (3), 197–216.
BOE (2007) Ley 25/2007, de 18 de octubre, de conservación de datos relativos a las comunicaciones electrónicas y a las redes públicas de comunicaciones. Law No. 251, Friday 19 October (pp. 42517–42523). Madrid: Gobierno de España, Ministerio de la Presidencia.

Boix i Fuster, E. (1993) *Triar no és trair. Identitat i llengua en els joves de Barcelona*. Barcelona: Edicions 62.
Boix i Fuster, E. and Vila i Moreno, F.X. (1998) *Sociolingüística de la llengua catalana*. Barcelona: Ariel.
Boldó Gaspá, M.D., Agustí, R., Muntada Balust, M., Nieto Trullàs, J. and Viñals, A. (1999) *La telefonía móvil en España*. Cuadernos Barcelona, Madrid, México. Escuela de Organización Industrial, Fundación Airtel: Mundi-Prensa.
BOP (2002) Terrassa. Aprovació definitiva de l'ordenança reguladora dels establiments de serveis telefònics per a ús públic. Law No. 128, 29 May (pp. 45–46). Barcelona: Diputació de Barcelona.
BOP (2005) Rubí. Aprovació definitiva de l'ordenança reguladora dels establiments de serveis telefònics per a ús públic. Law No. 136, 9 July (pp. 52–53). Barcelona: Diputació de Barcelona.
BOP (2006) Montcada i Reixac. Aprovació definitiva de l'ordenança municipal reguladora dels locutoris telefònics. Law No. 217, 11 September (pp. 21–23). Barcelona: Diputació de Barcelona.
BOP (2008) Ripollet. Projecte d'establiment i reglament per a l'establiment de funcionament del servei públic municipal de telecentre. Law No. 287 (3), 29 November (pp. 119–120). Barcelona: Diputació de Barcelona.
Bourdieu, P. (1977) *Outline of a Theory of Practice*. Cambridge: Cambridge University Press.
Bourdieu, P. (1986) The forms of capital. In J.G. Richardson (ed.) *Handbook of Theory and Research for the Sociology of Education* (pp. 241–258). New York: Greenwood Press.
Bourdieu, P. (1990) *The Logic of Practice*. Stanford, CA: Stanford University Press.
Bourdieu, P. (1991) *Language and Symbolic Power*. Cambridge, MA: Harvard University Press.
Bourdieu, P. and Wacquant, L.J.D. (1994) *Per a una sociologia reflexiva*. Barcelona: Herder.
Boutet, J. (2006) A atividade do trabalho nas centrais de atendimento: um trabalho de linguagem. Working in call centres: A language work. *Revista Brasileira de Saúde Ocupacional* 31 (114), 73–82.
Boutet, J. (2008) *La vie verbale au travail. Des manufactures aux centres d'appels*. Toulouse: Octares Editions.
Boutet, J. (2012) Language workers: Emblematic figures of late capitalism. In A. Duchêne and M. Heller (eds) *Language in Late Capitalism: Pride and Profit* (pp. 207–229). New York: Routledge.
Bucholtz, M. (2000) The politics of transcription. *Journal of Pragmatics* 32, 1439–1465.
Bührig, K. and Meyer, B. (2004) Ad hoc-interpreting and achievement of communicative purposes in briefings for informed consent. In J. House and J. Rehbein (eds) *Multilingual Communication* (pp. 43–62). Amsterdam: John Benjamins.
Çağlar, A.S. (2001) Constraining metaphors and the transnationalisation of spaces in Berlin. *Journal of Ethnic and Migration Studies* 27 (4), 601–613.
Cameron, D. (2000) *Good to Talk? Living and Working in a Communication Culture*. London: Sage.
Canyelles, J.M. (2011) Les llengües com a part integrant de la responsabilitat social de les empreses. In M. Strubell and I. Marí (eds) *Mercat global i Mercat local: implicacions per al multilingüisme de l'empresa. Actes del Seminari del CUIMPB-CEL 2008* (pp. 123–145). Barcelona: Universitat Oberta de Catalunya.
Carrasco Carpio, C. and García Serrano, C. (2012) *Inmigración y mercado de trabajo. Informe 2011*. Secretaría General de Inmigración y Emigración. Ministerio de Empleo y Seguridad Social. Madrid: Subdirección General de Información.

Castells, M. (1999) *Information Technology, Globalization and Social Development*. UNRISD Discussion Paper 114. Online at http://www.unrisd.org (accessed 13 July 2013).
Castells, M. (2000 [1996]) *The Rise of the Network Society*. Oxford: Blackwell.
Castells, M. (ed.) (2004) *The Network Society: A Cross-Cultural Perspective*. Cheltenham: Edward Elgar Publishing.
Castells, M. (2008) Afterword. In J.E. Katz (ed.) *Handbook of Mobile Communication Studies* (pp. 447–451). Cambridge, MA: MIT Press.
Castells, M. (2009) *Communication Power*. Oxford: Oxford University Press.
Castells, M. (2012) *Networks of Outrage and Hope: Social Movements in the Internet Age*. Cambridge: Polity Press.
Castells, M. and Portes, A. (1989) World underneath: The origins, dynamics, and effects of the informal economy. In A. Portes, M. Castells and L.A. Benton (eds) *The Informal Economy: Studies in Advanced and Less Developed Countries* (pp. 11–37). Baltimore, MD: Johns Hopkins University Press.
Castells, M., Tubella, I., Sancho, T., Díaz de Isla, M.I. and Wellman, B. (2004) Social structure, cultural identity, and personal autonomy in the practice of the internet: The network society in Catalonia. In M. Castells (ed.) *The Network Society: A Cross-Cultural Perspective* (pp. 233–248). Cheltenham: Edward Elgar Publishing.
Castells, M., Fernández-Ardèvol, M., Qiu, J.L. and Sey, A. (2007a) *Mobile Communication and Society: A Global Perspective*. Cambridge, MA: MIT Press.
Castells, M., Tubella, I., Sancho, T. and Roca, M. (2007b) *La transició a la societat xarxa*. Barcelona: Ariel.
Catalunya Press (2012) La Policia Nacional desmantella una xarxa de locutoris que blanquejava diners de la droga, 27 December. Online at http://www.catalunyapress.cat/cat/notices/2012/12/-la-policia-nacional-desmantella-una-xarxa-de-locutoris-que-blanquejava-diners-de-la-droga-74539.php (accessed 15 February 2013).
Cebrián de Miguel, J.A. and Bodega Fernández, M.I. (2002) El negocio étnico, nueva fórmula del comercio en el casco antiguo de Madrid. El caso de Lavapiés. *Estudios geográficos* 63 (248/249), 559–580.
Chick, J.K. (1996) Safe-talk: Collusion in apartheid education. In H. Coleman (ed.) *Society and the Language Classroom* (pp. 21–39). Cambridge: Cambridge University Press.
Chipchase, J. (2008) Reducing illiteracy as a barrier to mobile communication. In J.E. Katz (ed.) *Handbook of Mobile Communication Studies* (pp. 79–89). Cambridge, MA: MIT Press.
CiberP@ís (2007) El africano llama cada día, mientras que el latino lo hace en fin de semana. *El País Semanal* 477, 1–6.
CMT (2011) *Informe económico sectorial. Comunicaciones móviles*. Online at http://informecmt.cmt.es/docs/Informe%20economico%20sectorial/Comunicaciones%20moviles%20CMT%202011.pdf (accessed 1 July 2013).
Codó, E. (2008) *Immigration and Bureaucratic Control: Language Practices in the Public Administration*. Berlin: Mouton de Gruyter.
Codó, E. (2012) From Africa to Catalonia: Notes on a sociolinguistics of youth, migration and English. In S. Martín Alegre, M.G. Moyer, E. Pladevall and S. Tubau (eds) *At a Time of Crisis: English and American Studies in Spain* (pp. 308–316). AEDEAN Conference Proceedings. Departament de Filologia Anglesa i de Germanística, Universitat Autònoma de Barcelona. Online at http://www.aedean.org (accessed 13 July 2013).
Codó, E. and Garrido, M.R. (2010) Ideologies and practices of multilingualism in bureaucratic and legal advice encounters. In R. Márquez-Reiter and L. Martín-Rojo (eds) *Sociolinguistic and Pragmatic Aspects of Institutional Discourse: Service Encounters in*

Multilingual and Multicultural Contexts, special issue of *Sociolinguistic Studies* 4 (2), 297–332.
Codó, E., Patiño, A. and Unamuno, V. (2012) Hacer sociolingüística etnográfica en un mundo cambiante. Retos y aportaciones desde la perspectiva hispana. In E. Codó, A. Patiño and V. Unamuno (eds) *La sociolingüística con perspectiva etnográfica en el mundo hispano: nuevos contextos, nuevas realidades, nuevas aproximaciones*, special issue of *Spanish in Context* 9 (2), 167–190.
Collins, J. (2005) Afterword: 'Story, place and encounter'. In M. Baynham and A. De Fina (eds) *Dislocations/Relocations: Narratives of Displacement* (pp. 242–252). Manchester: Encounters (St Jerome).
Confederació de Comerç de Catalunya (2013) *Guia Bàsica del Locutori*. Barcelona: Confederació de Comerç de Catalunya, Generalitat de Catalunya, Diputació de Barcelona, Àrea de Desenvolupament Econòmic i Ocupació. Online at http://www.confecom.cat/serveiling/wp-content/uploads/2008/12/guia-locutori-castellano.pdf (accessed 30 January 2013).
Connectta't (2008) *Connectta't a les telecomunicacions de Catalunya*, issue no. 18. Barcelona: Collegi oficial d'enginyers tècnics de telecomunicació a Catalunya.
Consumer (2004) La competencia de los locutorios obliga a Telefónica a bajar sus tarifas internacionales hasta un 45%, 19 April. Online at http://www.consumer.es/web/es/economia_domestica/2004/04/19/98817.php (accessed 28 January 2009).
Copevo (2011) Taxa d'atur del Vallès Occidental. Setembre 2011. Consorci per l'ocupació i la promoció econòmica del Vallès Occidental. Consorci d'Ocupació del Vallès Occidental. Online at http://www.google.es/url?sa=t&rct=j&q=&esrc=s&source=web&cd=1&ved=0CDAQFjAA&url=http%3A%2F%2Fwww.consorciocupaciovalles.cat%2Fcopevoc%2Frecursos%2Frecursos%2Fv2info_atur_setembre_2011.pps&ei=AF_BUtnVNsT70gX_q4HwCg&usg=AFQjCNGmep4Z41ysJLTU8o0syRcSpzAJUw&sig2=r-zy50B7NkEAwJksBxpv4Q&bvm=bv.58187178,d.d2k (accessed 30 January 2013).
Corominas, J. (2008) *Breve diccionario etimológico de la lengua castellana*. Madrid: Gredos.
Corona, V., Nussbaum, L. and Unamuno, V. (2013) The emergence of new linguistic repertoires among Barcelona's youth of Latin American origin. In *Catalan in the Twenty-First Century*, special issue of *International Journal of Bilingual Education and Bilingualism* 16 (2), 182–194.
Coupland, N. (2003) Introduction: Sociolinguistics and globalisation. *Journal of Sociolinguistics* 7 (4), 465–472.
Coupland, N. (ed.) (2010) *The Handbook of Language and Globalization*. Oxford: Wiley-Blackwell.
Dana, L. (ed.) (2007) *Handbook of Research on Ethnic Minority Entrepreneurship: A Co-evolutionary View on Resource Management*. Cheltenham: Edward Elgar Publishing.
Danet, B. and Herring, S.C. (2007) Introduction: Welcome to the multilingual internet. In B. Danet and S.C. Herring (eds) *The Multilingual Internet: Language, Culture, and Communication Online* (pp. 1–39). Oxford: Oxford University Press.
Da Silva, E., McLaughlin, M. and Richards, M. (2007) Bilingualism and the globalised new economy: The commodification of language and identity. In M. Heller (ed.) *Bilingualism: A Social Approach* (pp. 183–206). London: Palgrave Macmillan.
De Fina, A. and Georgakopoulou, A. (2008) Analyzing narratives as practices. *Qualitative Research* 8 (3), 379–387.
Del Valle, J. (2006) US Latinos, *la hispanofonía*, and the language ideologies of high modernity. In C. Mar-Molinero and M. Stewart (eds) *Globalization and Language in*

the Spanish-Speaking World: Macro and Micro Perspectives (pp. 27–46). Houndmills: Palgrave Macmillan.
Del Valle, J. (2007) Embracing diversity for the sake of unity: Linguistic hegemony and the pursuit of total Spanish. In A. Duchêne and M. Heller (eds) *Discourses of Endangerment: Interest and Ideology in the Defence of Language* (pp. 242–267). London: Continuum.
Del Valle, J. and Gabriel-Stheeman, L. (2002) *The Battle Over Spanish Between 1800 and 2000*. London: Routledge.
Departament d'Ensenyament (2012) *Els plans educatius d'entorn. Document marc dels plans educatius d'entorn*. Barcelona: Generalitat de Catalunya. Online at http://www20.gencat.cat/docs/Educacio/Home/Arees_actuacio/innovacio_educativa/Plans%20educatius%20entorn/doc_marc_pee_cat.pdf (accessed 30 January 2013).
Diario de León (2004) La policía encontró huellas de Zougam la misma tarde del 11-M, 9 May. Online at http://www.diariodeleon.es/noticias/espana/policia-encontro-huellas-zougam-misma-tarde-11-m-cinco-primeros-sospechosos-fueron-detenidos-hora-media_136509.html (accessed 29 October 2009).
Díaz, B. (2004) *'Y así nos entendemos'. Lenguas y comunicación en la emigración. El barrio de San Francisco de Bilbao*. Madrid: Traficantes de Sueños; Bilbao: Likiniano Elkartea.
Diminescu, D. (2008) The connected migrant: An epistemological manifesto. *Social Science Information* 47 (4), 565–579.
Donner, J. (2008a) Shrinking Fourth World? Mobiles, development, and inclusion. In J.E. Katz (ed.) *Handbook of Mobile Communication Studies* (pp. 29–42). Cambridge, MA: MIT Press.
Donner, J. (2008b) Research approaches to mobile use in the developing world: A review of the literature. *Information Society* 24, 140–159.
Duchêne, A. (2009) Marketing, management and performance: Multilingualism as commodity in a tourist call centre. *Language Policy* 8, 27–50.
Duchêne, A. (2011) Néolibéralisme, inégalités sociales et plurilinguisme: L'exploitation des ressources langagières et des locuteurs. *Langage et Société* 136, 81–106.
Duchêne, A., Moyer, M.G. and Roberts, C. (2013) Introduction: Recasting institutions and work in multilingual and transnational spaces. In A. Duchêne, M.G. Moyer and C. Roberts (eds) *Language, Migration and Social Inequalities: A Critical Sociolinguistic Perspective on Institutions and Work* (pp. 1–21). Bristol: Multilingual Matters.
Edjabe, N. and Pieterse, E. (eds) (2011) *African Cities Reader II: Mobilities and Fixtures*. South Africa: Chimurenga and the African Centre for Cities.
ELAN.cat (2006) Són prou multilingües les empreses catalanes? UOC, Linguamón and Generalitat de Catalunya. Online at http://www10.gencat.cat/casa_llengues/binaris/opuscle_ELAN_tcm302-122718.pdf (accessed 30 January 2013).
Elias Boada, J. (2011) *Immigració i mercat laboral: Abans i després de la recessió*. Documents d'Economia No. 20. Barcelona: La Caixa, Estudis i Anàlisi Econòmica – Edicions 62.
El 9 Nou (2012) Un detingut en una inspecció a locutoris de Vic, 27 January. Online at http://www.el9nou.cat/tvcarta2_v_4_13/26/3872 (accessed 15 February 2013).
El País (1981) Sólo hay dos locutorios telefónicos para conferencias de larga distancia, 1 October. Online at http://elpais.com/diario/1981/10/01/madrid/370787061_850215.html (accessed 28 January 2009).
El País (1990) Desbarajuste en Barajas, 14 August. Online at http://elpais.com/diario/1990/08/14/opinion/650584806_850215.html (accessed 9 September 2008).
El País (1997) Telefonía 'de rebajas', 25 January. Online at http://elpais.com/diario/1997/01/25/madrid/854195064_850215.html (accessed 9 September 2009).

El País (2000) Vic Telehome controla la mitad de los 800 locutorios para inmigrantes, 11 June. Online at http://elpais.com/diario/2000/06/11/economia/960674403_850215.html (accessed 9 September 2009).

El País (2003) La CMT examina hoy una multa a Telefónica por su oferta a locutorios, 10 July. Online at http://elpais.com/diario/2003/07/10/economia/1057788010_850215.html (accessed 9 September 2008).

El País (2009) ¿Usas tarjeta prepago de móvil? ¡Regístrala!, 4 March. Online at http://elpais.com/elpais/2009/03/04/actualidad/1236158228_850215.html (accessed 1 June 2009).

El País (2013) Locutorios bajo el paraguas gremial, 29 May. Online at http://ccaa.elpais.com/ccaa/2013/05/28/catalunya/1369768416_324506.html (accessed 30 May 2013).

El Punt (2010) Detingudes tres presones per tràfic de drogues en un locutori, 5 February. Online at http://www.elpuntavui.cat/noticia/article/24-puntdivers/4-divers/133117-detingudes-tres-persones-per-trafic-de-drogues-en-un-locutori.html (accessed 15 February 2013).

Extra, G., Spotti, M. and Van Avermaet, P. (eds) (2009) *Language Testing, Migration and Citizenship. Cross-National Perspectives on Integration Regimes*. London: Continuum.

Fabà Prats, A. (2012) L'evolució dels usos professionals a Catalunya del 1997 al 2008. *Treballs de Sociolingüística Catalana* 22, 41–58.

Fairclough, N. (2006) *Language and Globalization*. London: Routledge.

Farlex (2012) Mogul. *The Free Dictionary*. Online at http://www.thefreedictionary.com/Mogul (accessed 13 July 2013).

Feliu, J., Peñaranda-Cólera, M.C. and Gil-Juárez, A. (2012) Comunidades imaginadas: Nacionalismo banal en los locutorios de Barcelona. *AIBR Revista de Antropología Iberoamericana* 7 (2), 197–224.

Ferguson, C.A. (1975) Toward a characterization of English foreigner talk. *Anthropological Linguistics* 17 (1), 1–14.

Fischer, C.S. (1992) *America Calling: A Social History of the Telephone to 1940*. Berkeley, CA: University of California Press.

FOBSIC and Idescat (2010) *Enquesta sobre l'equipament i l'ús de les Tecnologies de la Informació i la Comunicació (TIC) a les llars de Catalunya (2010). Volume II. Usos TIC individuals.* Fundació Observatori per la Societat de la Informació a Catalunya and Institut d'Estadística de Catalunya. Online at http://www.idescat.cat/pub/?id=ticll10 (accessed 30 January 2013).

FOBSIC and Plataforma per la Llengua (2007) *Estudi sobre la telefonia mòbil a Catalunya. Estudi sobre la presència de la llengua catalana en el sector de la telefonia mòbil, des del punt de vista dels usuaris i usuàries*. Plataforma per la Llengua and Fundació Observatori per la Societat de la Informació a Catalunya and Institut d'Estadística de Catalunya. Online at http://www.fobsic.cat (accessed 28 January 2009).

Fontela, E. and Guzmán, J. (2000) *La telefonía móvil en España II: Efectos sobre la productividad de las empresas*. Madrid: Fundación Airtel.

Fortunati, L. (2002) Italy: Stereotypes, true and false. In J.E. Katz (ed.) *Handbook of Mobile Communication Studies* (pp. 42–62). Cambridge, MA: MIT Press.

Foucault, M. (1979) *Discipline and Punish: The Birth of the Prison*. New York: Vintage Books.

Frekko, S.E. (2013) Legitimacy and social class in Catalan language education for adults. In *Catalan in the Twenty-First Century*, special issue of *International Journal of Bilingual Education and Bilingualism* 16 (2), 164–176.

Gal, S. (2001) Language, gender, and power: An anthropological review. In S. Duranti (ed.) *Linguistic Anthropology: A Reader* (pp. 420–430). Oxford: Blackwell.

Gal, S. (2006) Migration, minorities and multilingualism: Language ideologies in Europe. In C. Mar-Molinero and P. Stevenson (eds) *Language Ideologies, Policies and Practices* (pp. 13–27). Basingstoke: Palgrave.
Galeano, E. (1993) *El Libro de los Abrazos*. Madrid: Siglo XXI.
García Delgado, J.L., Alonso, J.A. and Jiménez, J.C. (2007) *Economía del Español. Una introducción*. Barcelona: Ariel; Madrid: Fundación Telefónica.
Garrido, M.R. (2010) 'If you slept in Catalunya you know that here it's a paradise': Multilingual practices and ideologies in a residential project for migrants. Unpublished MA thesis, Universitat Autònoma de Barcelona.
Garzón, L. (2011) Second generation Argentineans: Between identity and mobilization. In A. Alarcón and L. Garzón (eds) *Language, Migration and Social Mobility in Catalonia*. (pp. 33–62). Leiden: Brill.
Garzón, L. (2012) Globalization, Latin American migration and Catalan: Closing the ring. *Sustainability* 4, 2498–2512.
Geertz, C. (1973) *The Interpretation of Cultures: Selected Essays*. New York: Basic Books.
Generalitat de Catalunya (2006) *Estatut d'Autonomia de Catalunya*. Barcelona: BIGSA Indústria Gràfica.
Generalitat de Catalunya (2007) Pla per la llengua i la cohesió social. Generalitat de Catalunya. Departament d'Educació. Online at http://www.xtec.cat/lic/intro/documenta/pil.pdf (accessed 30 January 2013).
Generalitat de Catalunya (2008) National Agreement on Immigration. An agreement to live together. Departament de Benestar Social i Família. Online at. http://www20.gencat.cat/docs/dasc/03Ambits%20tematics/05Immigracio/03Politiquesplansactuac io/02pactenacionalimmigracio/02continguts/Pdfs/Document_final_PNI_angles.pdf (accessed 30 January 2013).
Generalitat de Catalunya (2009) Informe de resultats de les enquestes d'usos lingüístics a les empreses de serveis. 2004–2007. Biblioteca tècnica de política lingüística. Dades i estudis. Col·legi de Politòlegs i Sociòlegs de Catalunya. Generalitat de Catalunya. Departament de la Vicepresidència. Secretaria de Política Lingüística. Online at http://www20.gencat.cat/docs/Llengcat/Documents/Publicacions/BTPL/ arxius/2_EULE. pdf (accessed 30 January 2013).
Generalitat de Catalunya (2011a) El Departament d'Empresa i Ocupació presenta la primera 'Diagnosi del mercat de treball a Catalunya'. Departament d'Empresa i Ocupació and Servei d'Ocupació de Catalunya. Comunicat de premsa, 25 July. Online at http://premsa.gencat.cat/pres_fsvp/AppJava/notapremsavw/detall. do?id=117091 (accessed 30 January 2013).
Generalitat de Catalunya (2011b) Comarques de Catalunya. Generalitat de Catalunya. Online at http://www20.gencat.cat/portal/site/PalauRobert/menuitem.698fb9529 405d0f972623b10b0c0e1a0/?vgnextoid=8ef8bbae87492110VgnVCM1000000b0c1e 0aRCRD (accessed 30 January 2013).
Giddens, A. (1984) *The Constitution of Society: Outline of the Theory of Structuration*. Cambridge: Polity Press.
Giddens, A. (1991) *Modernity and Self-Identity: Self and Society in the Late Modern Age*. Cambridge: Polity Press.
Glick Schiller, N. and Çağlar, A. (2013) Locating migrant pathways of economic emplacement: Thinking beyond the ethnic lens. *Ethnicities* 13 (4) 494–514. Online at http://www.academia.edu/2577039/Locating_migrant_pathways_of_economic_ emplacement_Thinking_beyond_the_ethnic_lens (accessed 13 July 2013).
Goffman, E. (1959) *The Presentation of Self in Everyday Life*. Garden City, NY: Doubleday.

Goffman, E. (1981) *Forms of Talk*. Philadelphia, PA: University of Pennsylvania Press.
Goggin, G. (2006) *Cell Phone Culture: Mobile Technology in Everyday Life*. London: Routledge.
Gramsci, A. (1999 [1971]) *Selections from the Prison Notebooks* (eds Q. Hoare and G. Nowell Smith). London: Elecbook. Online at http://www.walkingbutterfly.com/wp-content/uploads/2010/12/gramsci-prison-notebooks-vol1.pdf (accessed 13 July 2013).
GSM Spain (2009) Los morosos en España superaron los 2,7 millones a cierre de 2008, 2 February. Online at http://www.gsmspain.com/foros/h716679_Off-topic-Economia-finanzas_morosos-Espana-superaron-millones-cierre-2008.html (accessed 2 July 2013).
Guillén, M.F. (2005) *The Rise of Spanish Multinationals: European Business in the Global Economy*. Cambridge: Cambridge University Press.
Gumperz, J.J. (1982) *Discourse Strategies*. Cambridge: Cambridge University Press.
Gumperz, J.J. (1986) Interactional sociolinguistics in the study of schooling. In J. Cook-Gumperz (ed.) *The Social Construction of Literacy* (pp. 45–68). Cambridge: Cambridge University Press.
Hannam, K., Sheller, M. and Urry, J. (2006) Editorial: Mobilities, immobilities and moorings. *Mobilities* 1 (1), 1–22.
Harvey, D. (2003) *The New Imperialism*. Oxford: Oxford University Press.
Harvey, D. (2005) *A Brief History of Neoliberalism*. Oxford: Oxford University Press.
Harvey, D. (2010) *The Enigma of Capital and the Crises of Capitalism*. Oxford: Oxford University Press.
Heller, M. (2001) Undoing the macro/micro dichotomy: Ideology and categorization in a linguistic minority school. In N. Coupland, S. Sarangi and C.N. Candlin (eds) *Sociolinguistics and Social Theory* (pp. 212–234). London: Longman.
Heller, M. (2003) Globalization, the new economy, and the commodification of language and identity. *Journal of Sociolinguistics* 7 (4), 473–492.
Heller, M. (2006 [1999]) *Linguistic Minorities and Modernity: A Sociolinguistic Ethnography*. London: Continuum.
Heller, M. (2007a) Doing ethnography. In L. Wei and M.G. Moyer (eds) *The Blackwell Handbook of Research: Methods in Bilingualism and Multilingualism* (pp. 249–262). Oxford: Blackwell.
Heller, M. (2007b) Distributed knowledge, distributed power: A sociolinguistics of structuration. *Text and Talk* 27 (5–6), 633–653.
Heller, M. (2007c) Bilingualism as ideology and practice. In M. Heller (ed.) *Bilingualism: A Social Approach* (pp. 1–22). London: Palgrave Macmillan.
Heller, M. (2010a) *Paths to Post-Nationalism. A Critical Ethnography of Language and Identity*. New York: Oxford University Press.
Heller, M. (2010b) Language as resource in the globalised new economy. In N. Coupland (ed.) *Handbook of Language and Globalization* (pp. 349–365). Oxford: Blackwell.
Heller, M. and Boutet, J. (2006) Vers de nouvelles formes de pouvoir langagier? Langue(s) et identité dans la nouvelle économie. *Langage et Société* 118, 5–16.
Heller, M. and Duchêne, A. (2012) Pride and profit: Changing discourses of language, capital and nation-state. In A. Duchêne and M. Heller (eds) *Language in Late Capitalism: Pride and Profit* (pp. 1–21). New York: Routledge.
Heller, M. and Martin-Jones, M. (eds) (2001) *Voices of Authority: Education and Linguistic Difference*. Westport, CT: Ablex.
Hernández-Carr, A. (2011) Second generation Moroccan migrants in Catalonia: Language, employment and inclusion. In A. Alarcón and L. Garzón (eds) *Language, Migration and Social Mobility in Catalonia* (pp. 95–122). Leiden: Brill.

References   191

Hogan-Brun, G., Mar-Molinero, C. and Stevenson, P. (2009) Testing regimes: Introducing cross-national perspectives on language, migration and citizenship. In G. Hogan-Brun, C. Mar-Molinero and P. Stevenson (eds) *Discourses on Language and Integration: Critical Perspectives on Language Testing Regimes in Europe* (pp. 1–14). Amsterdam: John Benjamins.
Hondagneu-Sotelo, P. and Avila, E. (1997) 'I'm here but I'm there': The meaning of Latina transnational motherhood. *Gender and Society* 11 (5), 548–571.
Horst, H.A. and Miller, D. (2006) *The Cell Phone: An Anthropology of Communication*. Oxford: Berg.
Howard, P.N. (2002) Network ethnography and the hypermedia organization: New media, new organizations, new methods. *New Media and Society* 4 (4), 550–574.
Huguet, Á. (2007) Language use and language attitudes in Catalonia. In D. Lasagabaster and Á. Huguet (eds) *Multilingualism in European Bilingual Contexts: Language Use and Attitudes* (pp. 17–39). Clevedon: Multilingual Matters.
Idescat (2001) Estadístiques de població. Lloc de naixement. Online at http://www.idescat.cat/pub/?id=aec&n=257&t=2001&x=6&y=4 (accessed 13 July 2013).
Idescat (2010a) Població estrangera. Evolució a Catalunya. Institut d'Estadística de Catalunya. Online at http://www.idescat.cat/poblacioestrangera/?b=0 (accessed 13 July 2013).
Idescat (2010b) Moviments migratoris 2009. Online at http://www.idescat.cat/novetats/?id=825 (accessed 13 July 2013).
Idescat (2010c) Vallès Occidental, població per lloc de naixement. Institut d'Estadística de Catalunya. Online at http://www.idescat.cat/pub/?id=aec&n=257&t=2011&x=7&y=9 (accessed 13 July 2013).
Idescat (2011) Població segons nacionalitat. Institut d'Estadística de Catalunya. Online at http://www.idescat.cat/territ/basicterr?TC=5&V0=3&V1=0&V3=477&V4=478&ALLINFO=TRUE&PARENT=1&CTX=B (accessed 13 July 2013).
IFEMA (2008) Integra Madrid. III Feria de Productos y Servicios para Inmigrantes. Online at http://www.ifema.es/ferias/integramadrid/default.html (accessed 13 July 2013).
Inda, J.X. (2006) *Targeting Immigrants: Government, Technology and Ethics*. Malden, MA: Blackwell.
Inda, J.X. and Rosaldo, R. (eds) (2002) *The Anthropology of Globalization: A Reader*. Malden, MA: Blackwell.
INE (2011) Avance del padrón municipal a 1 de enero de 2011. Instituto Nacional de Estadística. Online at http://www.ine.es/prensa/np648.pdf (accessed 13 July 2013).
INE (2012) Encuesta de población activa. Tasas de actividad y paro. Instituto Nacional de Estadística. Online at http://www.ine.es (accessed 13 July 2013).
Íñiguez-Rueda, L., Martínez, L.M., Muñoz-Justicia, J., Peñaranda-Cólera, M.C. and Vitores González, A. (2012) Telecenters as association stations: The role of information and communication technologies in migratory processes. *Migraciones Internacionales* (6) 4, 75–105.
Irvine, J.T. and Gal, S. (2000) Language ideology and linguistic differentiation. In P.V. Kroskrity (ed.) *Regimes of Language: Ideologies, Polities, and Identities* (pp. 35–83). Santa Fe, NM: School of American Research Press and James Currey Ltd.
ITU (2011) About ITU. Online at http://www.itu.int/en/about/Pages/default.aspx (accessed 13 July 2013).
ITU (2012a) *Measuring the Information Society*. Geneva: International Telecommunication Union.

ITU (2012b) *The Little Data Book on Information and Communication Technology.* Washington DC: World Bank and International Telecommunications Union. Online at http://www.itu.int/ITU-D/ict/publications/material/LDB_ICT_2012.pdf (accessed 13 July 2013).

Jacquemet, M. (2005) Transidiomatic practices: Language and power in the age of globalization. *Language and Communication* 25 (3), 257–277.

Jacquemet, M. (2010) Language and transnational spaces. In P. Auer and J.E. Schmidt (eds) *Language and Space: An International Handbook of Linguistic Variation* (pp. 50–69). New York: Mouton de Gruyter.

Jaffe, A. (2000) Non-standard orthography and non-standard speech. *Journal of Sociolinguistics* 4 (4), 497–513.

Jaffe, A. (2007) Minority language movements. In M. Heller (ed.) *Bilingualism: A Social Approach* (pp. 50–70). London: Palgrave Macmillan.

Jaffe, A. (2012) Collaborative practice, linguistic anthropological enquiry and the mediation between researcher and practitioner discourses. In S. Gardner and M. Martin-Jones (eds) *Multilingualism, Discourse and Ethnography* (pp. 334–352). New York: Routledge.

Jaffe, A. and Walton, S. (2000) The voices people read: Orthography and the representation of non-standard speech. *Journal of Sociolinguistics* 4 (4), 561–587.

Jaworski, A. and Thurlow, C. (2010) Introducing semiotic landscapes. In A. Jaworski and C. Thurlow (eds) *Semiotic Landscapes: Language, Image, Space* (pp. 1–40). London: Continuum.

Joaquín López, J. (2006) El locutorio del pakistaní. Anecdotario blog entry dated 11 April. Online at http://www.anecdotario.net (accessed 2 February 2012).

Jørgensen, J.N., Rindler-Schjerve, R. and Vetter, E. (2012) Polylingualism, multilingualism, plurilingualism. A toolkit for transnational communication in Europe. Online at http://www.toolkit-online.eu/docs/polylingualism.html (accessed 13 July 2013).

Kahn, R. and Kellner, D. (2005) Reconstructing technoliteracy: A multiple literacies approach. *E-learning* 2 (3), 399–423.

Kamwangamalu, N.M. (2008) Language policy, vernacular education and language economics in postcolonial Africa. In P.K.W. Tan and R. Rubdy (eds) *Language as Commodity. Global Structures, Local Marketplaces* (pp. 171–186). London: Continuum.

Katz, J.E. (ed.) (2008) *Handbook of Mobile Communication Studies.* Cambridge, MA: MIT Press.

Kelly-Holmes, H. (2005) *Advertising as Multilingual Communication.* London: Palgrave Macmillan.

Kelly-Holmes, H. (2010) Markets and languages: Sociolinguistic perspectives. In H. Kelly-Holmes and G. Mautner (eds) *Language and the Market* (pp. 20–29). London: Palgrave Macmillan.

Kelly-Holmes, H. and Mautner, G. (eds) (2010) *Language and the Market.* London: Palgrave Macmillan.

Kohler Riessman, C. (2000) Stigma and everyday resistance practices: Childless women in South India. In *Emergent and Reconfigured Forms of Family Life*, special issue of *Gender and Society* 14 (1), 111–135.

Kostova Karaboytcheva, M. (2006) Una evaluación del último proceso de regularización de trabajadores extranjeros en España (febrero-mayo de 2005). Un año después. Área: Demografía, Población y Migraciones Internacionales. DT No. 15/2006. Madrid: Real Instituto Elcano. Online at http://www.realinstitutoelcano.org/documentos/252/252_kostova_regularizacion_extranjeros_espana.pdf (accessed 13 July 2013).

Kress, G. and van Leeuwen, T. (2006 [1996]) *Reading Images: The Grammar of Visual Design*. London: Routledge.
Kroskrity, P.V. (ed.) (2000) *Regimes of Language: Ideologies, Polities, and Identities*. Santa Fe, NM: School of American Research Press and James Currey Ltd.
Krzyzanowski, M. and Wodak, R. (2011) Political strategies and language policies: The European Union Lisbon strategy and its implications for the EU's language and multilingualism policy. *Language Policy* 10 (2), 115–136.
Lam, E.W.S. and Rosario-Ramos, E. (2009) Multilingual literacies in transnational digitally mediated contexts: An exploratory study of immigrant teens in the United States. *Language and Education* 23 (2), 171–190.
Lam, E.W.S. and Warriner, D. (2012) Transnationalism and literacy: Investigating the mobility of peoples, languages, texts, and practices in the context of migration. *Reading Research Quarterly* 47 (2), 191–215.
Larsen, J. and Hviid Jacobsen, M. (2009) Metaphors of mobility: Inequality on the move. In T. Ohnmacht, H. Maksim and M. Max Bergman (eds) *Mobilities and Inequality* (pp. 75–96). Surrey: Ashgate.
Lewicki, R.J., McAllister, D.J. and Bies, R.J. (1998) Trust and distrust: New relationships and realities. *Academy of Management Review* 23 (3), 438–458.
Lewin Tapia, J.G. (2004) Cuando el océano ya no es la última frontera: Una relación a (muy) larga distancia a través de las TICs. *Athenea Digital* 6, 1–12. Online at http://antalya.uab.es/athenea/num6/lewin.pdf (accessed 13 July 2013).
Lewis, M.P., Simons, G.F. and Fennig, C.D. (eds) (2011) The languages of Pakistan. In *Ethnologue: Languages of the World* (17th edition). Dallas, TX: SIL International. Online at http://www.ethnologue.com/country/PK (accessed 13 July 2013).
L'Hiperbòlic (2008) El telèfon mòbil supera el fix. *L'Hiperbòlic* issue 61, pp. 12–13 (periodical published in Palma by L'Hiperbòlic Edicions).
Light, I.H. (1972) *Ethnic Enterprise in America: Business and Welfare Among Chinese, Japanese and Blacks*. Berkeley, CA: University of California Press.
Lindgren, M., Jedbratt, J. and Svensson, E. (2002) *Beyond Mobile: People, Communications and Marketing in a Mobilized World*. New York: Palgrave.
Ling, R. (2004) *The Mobile Connection: The Cell Phone's Impact on Society*. San Francisco, CA: Morgan Kaufmann.
Linguamón (2010) Per saber-ne més sobre les llengües de la immigració a Catalunya. Linguamón. Casa de les Llengües. Generalitat de Catalunya. Online at http://www10.gencat.cat/casa_llengues/AppJava/ca/diversitat/diversitat/llengues_immigracio.jsp (accessed 13 July 2013).
LIPPS Group (2000) The LIDES coding manual: A document for preparing and analysing language interaction data. *International Journal of Bilingualism* 4 (2), 131–270.
Llurda, E. (2009) Attitudes towards English as an international language. In F. Sharifian (ed.) *English as an International Language: Perspectives and Pedagogical Issues* (pp. 119–134). Bristol: Multilingual Matters.
Luna René, J., Muñoz J. and Rodríguez, J.A. (2007) A social network analysis of discourse about technoscience in recent immigrants from the majority world to Spain. Paper presented at the Applications of Social Network Analysis Conference, 12–15 September, Institute of Mass Communication and Research, University of Zurich, Switzerland. Online at http://www.friemel.com/asna/papers/ASNA2007_Paper_Luna-Munoz-Rodriguez.pdf (accessed 13 July 2013).
MacWhinney, B. (2000 [1991]) *The CHILDES Project: Tools for Analyzing Talk*. Mahwah, NJ: Lawrence Erlbaum.

Mansoon, S. (2004) The status and role of regional languages in higher education in Pakistan. *Journal of Multilingual and Multicultural Development* 25 (4), 333–353.
Marcus, G.E. (1995) Ethnography in/of the world system: The emergence of multi-sited ethnography. *Annual Review of Anthropology* 24, 95–117.
Marí, I. and Strubell, M. (2011) Pròleg. In M. Strubell and I. Marí (eds) *Mercat global i Mercat local: implicacions per al multilingüisme de l'empresa. Actes del Seminari del CUIMPB-CEL 2008* (pp. 13–14). Barcelona: Universitat Oberta de Catalunya.
Mar-Molinero, C. (2006) Forces of globalization in the Spanish-speaking world: Linguistic imperialism or grassroots adaptation. In C. Mar-Molinero and M. Stewart (eds) *Globalization and Language in the Spanish-Speaking World: Macro and Micro Perspectives* (pp. 8–26). Houndmills: Palgrave Macmillan.
Marshall, S. (2006) Spanish-speaking Latin Americans in Catalonia: Reflexivity and knowledgeability in constructions of Catalan. In C. Mar-Molinero and M. Stewart (eds) *Globalization and Language in the Spanish-Speaking World: Macro and Micro Perspectives* (pp. 158–177). Houndmills: Palgrave MacMillan.
Marshall, S. (2007) New Lation diaspora and new zones of language contact: A social constructionist analysis of Spanish-speaking Latin Americans in Catalonia. In J. Holmquist, A. Lorenzino and L. Sayahi (eds) *Selected Proceedings of the Third Workshop on Spanish Sociolinguistics* (pp. 150–161). Sommerville, MA: Cascadilla Proceedings Project. Online at http://www.lingref.com/cpp/wss/3/paper1536.pdf (accessed 13 July 2013).
Martin-Jones, M. and Jones, K. (2000) Multilingual literacies. In M. Martin-Jones and K. Jones (eds) *Multilingual Literacies: Reading and Writing Different Worlds* (pp. 1–15). Amsterdam: John Benjamins.
Martín Municio, A., Espasa, A., Girón, J. and Peña, D. (eds) (2003) *El valor económico de la lengua española*. Madrid: Espasa Calpe.
Martín-Rojo, L. (ed.) (2010) *Constructing Inequality in Multilingual Classrooms*. Berlin: Mouton de Gruyter.
Martín-Rojo, L. (2012) Paisajes lingüísticos de indignación. Prácticas comunicativas para tomar las plazas. *Anuari del Conflicte Social 2012*, 275–302.
Martín-Rojo, L. and Molina, C. (2012) Madrid multilingüe: Lengua pa' la citi. Online at http://web.uam.es/ss/Satellite/es/1242652961025/1242664605633/articul o/articulo/Madrid_multilingue:_Lenguas_pa%E2%80%99_la_city.htm (accessed 13 July 2013).
Metro (2008a) El éxito es móvil, 29 April (Noticias: 15. Barcelona: Metro News S.L.).
Metro (2008b) Advertisement by Lebara, 26 February (Publicitat: 17. Barcelona: Metro News S.L.).
Migracat (2011a) Tipus d'ingrés principal per lloc de naixement. Observatori de la Immigració a Catalunya. Fundació Jaume Bofill. Online at http://www.migracat.cat (accessed 13 July 2013).
Migracat (2011b) Evolució de l'afiliació a la Seguretat Social: Règim d'Autònoms, 2009. Observatori de la Immigració a Catalunya. Fundació Jaume Bofill. Online at http://www.migracat.cat (accessed 13 July 2013).
Ministerio del Interior (2007) '¡Identifícate!'. Plan de identificación obligatoria de usuarios de tarjetas telefónicas prepago. Ministerio del Interior. Gobierno de España. Online at http://www.interior.gob.es/DGRIS/Notas_Prensa/PDF_notas_de_prensa/2009/identificate_mir.pdf (accessed 13 July 2013).
Ministerio del Interior (2012) Lucha contra la inmigración ilegal. Balance 2011. Ministerio del Interior. Gobierno de España. Online at http://www.interior.gob.es/file/54/54239/54239.pdf (accessed 13 July 2013).

Miyata, K., Wellman, B. and Boase, J. (2005) The wired – and wireless – Japanese: Webphones, PCs and social networks. In R. Ling and P.E. Pedersen (eds) *Mobile Communications: Re-negotiation of the Social Sphere* (pp. 427–449). London: Springer.

Molins Pueyo, C. (2006) Use and misuse of information and communication technologies in education in Spain: Limits to change and cultural production. *Electronic Magazine of Multicultural Education* 8 (1), 1–15. Online at http://www.eastern.edu/publications/emme/2006spring/molins-pueyo.pdf (accessed 13 July 2013).

Moran-Taylor, M.J. (2008) When mothers and fathers migrate north: Caretakers, children, and child rearing in Guatemala. *Latin American Perspectives* 35 (4), 79–95.

Moreras, J. (2007) Iniciativas comerciales inmigrantes en un contexto urbano en transformación: El caso de Ciutat Vella (Barcelona). In J. Beltrán, L. Oso and N. Ribas (eds) *Empresariado étnico en España* (pp. 128–154). Observatorio Permanente de la Inmigración. Barcelona: Centre d'Estudis i Documentació Internacional de Barcelona. Madrid: Subdirección General de Información Administrativa y Publicaciones.

Movistar (2013) Tarifa Habla Internacional. Madrid: Telefónica S.A. Online at http://www.movistar.es/particulares/movil/tarifas-tarjeta/ficha/tarifa-habla-internacional (accessed 13 July 2013).

Moyer, M.G. (2010) The management of multilingualism in public, private and non-governmental institutions. *Sociolinguistic Studies* 4 (2), 267–296.

Moyer, M.G. (2011) What multilingualism? Agency and the unintended consequences of multilingual practices in a Barcelona health clinic. *Journal of Pragmatics* 43 (5), 1209–1221.

Moyer, M.G. and Martín-Rojo, L. (2007) Language, migration and citizenship: New challenges in the regulation of bilingualism. In M. Heller (ed.) *Bilingualism: A Social Approach* (pp. 137–160). London: Palgrave Macmillan.

Muehlmann, S. and Duchêne, A. (2007) Beyond the nation-state: International agencies as new sites of discourses on bilingualism. In M. Heller (ed.) *Bilingualism: A Social Approach* (pp. 96–110). London: Palgrave Macmillan.

Nach (2005) Tierra prometida. *Ars Magna – Miradas*. Madrid: Boa Musica.

Narotzky, S. and Smith, G. (2006) *Immediate Struggles: People, Power and Place in Rural Spain*. Berkeley, CA: University of California Press.

Newman, M., Patiño-Santos, A. and Trenchs-Parera, M. (2013) Linguistic reception of Latin American students in Catalonia and their responses to educational language policies. In *Catalan in the Twenty-First Century*, special issue of *International Journal of Bilingual Education and Bilingualism* 16 (2), 195–209.

Nielsen (2007) Ómnibus inmigración 2008. Una oportunidad para llegar a nuevos consumidores. Online at http://es.nielsen.com/site/documents/OmnibusInmigracion_PresentacionEstudio2008.pdf (accessed 13 July 2013).

Normann Jørgensen, J. and Juffermans, K. (2011) Superdiversity. A toolkit for transnational communication in Europe. Documents. Online at http://www.toolkit-online.eu/docs/superdiversity.html (accessed 13 July 2013).

Observatori d'Empresa i Ocupació (2012) *Butlletí de Població Estrangera i Mercat de Treball*. Generalitat de Catalunya, Departament d'Empresa i Ocupació.

Observatorio Permanente de la Inmigración (2010) *Trabajadores extranjeros en alta laboral en la Seguridad Social según sexo, provincia y régimen de Seguridad Social 31-12-2009. Anuario Estadístico de Inmigración 16/11/2010*. Secretaría General de Inmigración y Emigración, Ministerio de Empleo y Seguridad Social. Gobierno de España. http://extranjeros.empleo.gob.es/es/ObservatorioPermanenteInmigracion/Anuarios/index.html (accessed 13 July 2013).

Ochs, E. (1979) Transcription as theory. In E. Ochs and B.B. Schieffelin (eds) *Developmental Pragmatics* (pp. 43–72). New York: Academic Press.

O'Neill, P.D. (2003) The 'poor man's mobile telephone': Access versus possession to control the information gap in India. *Contemporary South Asia* 12 (1), 85–102.

ONTSI (2011) *Penetración del teléfono móvil en los hogares españoles 2002–2010*. Observatorio Nacional de las Telecomunicaciones y de la SI, Ministerio de Industria, Turismo y Comercio. Online at http://www.ontsi.red.es/ontsi/es/estudios-informes/xxx-oleada-del-panel-de-hogares-octubre-diciembre-2010-0 (accessed 13 July 2013).

Oso Casas, L. and Villares Varda, M. (2005) Mujeres inmigrantes Latinoamericanas y empresariado étnico; dominicanas en Madrid, argentinas y venezolanas en Galicia. *Revista Galega de Economía* 14 (1/2), 1–19.

Overa, R. (2008) Mobile traders and mobile phones in Ghana. In J.E. Katz (ed.) *Handbook of Mobile Communication Studies* (pp. 43–54). Cambridge, MA: MIT Press.

Paetsch, M. (1993) *Mobile Communications in the US and Europe: Regulations, Technology, and Markets*. Boston, MA: Artech House.

Páginas Amarillas (2012) Locutorios en la provincia de Barcelona. Online at http://www.paginasamarillas.es/search/locutorios (accessed 1 February 2012).

Pagliarin, S. (2012) Empresariado étnico y formación de enclaves comerciales: El papel de las redes sociales en el caso de la calle de Sant Pau en Barcelona. *Biblio 3W Revista Bibliográfica de Geografía y Ciencias Sociales* 17 (962). Online at http://www.ub.edu/geocrit/b3w-962.htm (accessed 13 July 2013).

Palfreyman, D. and Al Khalil, M. (2007) 'A funky language for teenzz to use': Representing Gulf Arabic in instant messaging. In B. Danet and S.C. Herring (eds) *The Multilingual Internet: Language, Culture, and Communication Online* (pp. 43–63). Oxford: Oxford University Press.

Panagakos, A.N. and Horst, H.A. (2006) Return to Cyberia: Technology and the social worlds of transnational migrants. *Global Networks* 6 (2), 109–124.

Paolillo, J.C. (1996) Language choice on soc.culture.punjab. *Electronic Journal of Communication* 6 (3). Online at http://www.cios.org/EJCPUBLIC/006/3/006312.HTML (accessed 13 July 2013).

ParaInmigrantes.info (2007) El 11 de octubre se inaugura Integra Madrid la mayor feria dirigida al colectivo inmigrante de España, 27 September. Online at http://www.parainmigrantes.info/el-11-de-octubre-se-inaugura-intregra-madrid-la-mayor-feria-dirigida-al-colectivo-inmigrante-de-espana (accessed 7 July 2009).

ParaInmigrantes.info (2009) Integra 2009: La mayor feria para inmigrantes de España, 19 November. Online at http://www.parainmigrantes.info (accessed 13 January 2012).

ParaInmigrantes.info (2011) La Feria de la Inmigración, que tendrá lugar en Madrid será el marco para este foro de opinión en el que intervendrán grandes expertos en el tema, 11 July. Online at http://www.parainmigrantes.info (accessed 13 January 2012).

Parella, S. and Cavalcanti, L. (2008) Aplicación de los campos sociales transnacionales en los estudios sobre migraciones. In C. Solé, S. Parella and L. Cavalcanti (eds) *Nuevos retos del transnacionalismo en el estudio de las migraciones* (pp. 217–243). Ministerio de Trabajo e Inmigración. Subdirección General de Información. Madrid: Grafo.

Parella Rubio, S. (2005) Estrategias de los comercios étnicos en Barcelona, España. *Política y Cultura* 23, 257–275.

Parlament de Catalunya (2007) *Constitució espanyola*. Barcelona: Forma Color S.A.

Patten, A. and Kymlicka, W. (2003) Introduction. Language rights and political theory: Context, issues, and approaches. In W. Kymlicka and A. Patten (eds) *Language Rights and Political Theory* (pp. 1–51). Oxford: Oxford University Press.

Pavez Soto, I. (2011) Social mobility and the Catalan language: Analysis of second generation migrants from Colombia. In A. Alarcón and L. Garzón (eds) *Language, Migration and Social Mobility in Catalonia* (pp. 63–94). Leiden: Brill.

Pavlenko, A. (2007) Autobiographic narratives as data in applied linguistics. *Applied Linguistics* 28 (2), 163–188.

Pecóud, A. (2000) Thinking and rethinking ethnic economies. *Diaspora: A Journal of Transnational Studies* 9 (3), 439–462.

Pedone, C. and Gil Araújo, S. (2008) Maternidades transnacionales entre América Latina y el Estado español. El impacto de las políticas migratorias en las estrategias de reagrupación familiar. In C. Solé, S. Parella and L. Cavalcanti (eds) *Nuevos retos del transnacionalismo en el estudio de las migraciones* (pp. 149–176). Ministerio de Trabajo e Inmigración. Subdirección General de Información. Madrid: Grafo.

Peñaranda Cólera, M.C. (2005) El locutorio como espacio social transnacional: Una mirada psicosocial 1. *Athenea Digital 8*. Online at http://psicologiasocial.uab.es (accessed 13 July 2013).

Peñaranda Cólera, M.C. (2011) Migrando en tiempos de globalización: Usos de tecnologías de la información y la comunicación en contextos migratorios transnacionales. In F.J. García Castaño and N. Kressova (eds) *Actas del I Congreso Internacional sobre migraciones en Andalucía* (pp. 2023–2032). Granada: Instituto de Migraciones.

Pennycook, A. (2007) *Global Englishes and Transcultural Flows*. London: Routledge.

Pennycook, A. (2010) *Language as a Local Practice*. Abingdon: Routledge.

Pennycook, A. (2012) *Language and Mobility: Unexpected Places*. Bristol: Multilingual Matters.

Piller, I. (2011) *Intercultural Communication: A Critical Introduction*. Edinburgh: Edinburgh University Press.

Plan Avanza (2012) Estrategia general 2011–2015. Online at http://www.planavanza.es (accessed 13 July 2013).

Plataforma per la Llengua, Consorci per a la Normalització Lingüística, Fundació Jaume Bofill, Direcció General de Política Lingüística de la Generalitat de Catalunya (2008) Si us plau, parla'm en català. Online at http://www.parlacatala.cat/index2.html (accessed 22 March 2013).

Plataforma per la Llengua and Consum Català (2011) *El Català a la Telefonia Mòbil: Quarta Enquesta sobre Usos Lingüístics en la Telefonia Mòbil*. Departament de Governació i Administracions Públiques, Generalitat de Catalunya. Online at http://www.plataforma-llengua.cat/media/assets/1981/Estudi_Telef_nia_m_bil_2011.pdf (accessed 13 July 2013).

Players4Players (2008) La Asociación de Locutorios y Cíbers Españoles firma un acuerdo con Microsoft para poder salvar al sector de la crisis, 2 November. Online at http://www.players4players.com/nota_de_prensa/3289/la-asociacion-de-locutorios-y-cibers-espanoles-firma-un-acuerdo-con-microsoft-para-poder-salvar-al-sector-de-la-crisis (accessed 28 January 2009).

Portes, A., Haller, W. and Guarnizo, L.E. (2001) *Transnational Entrepreneurs: The Emergence and Determinants of an Alternative Form of Immigrant Economic Adaptation*. Working Papers Transnational Communities Programme WPTC-01-05. Oxford: Oxford University Press. Online at http://www.transcomm.ox.ac.uk/working%20papers/WPTC-01-05%20Portes.pdf (accessed 13 July 2013).

Pratt, M.L. (1991) Arts of the contact zone. *Profession* 91, 33–40.

Puig, N., Álvaro, A. and Castro, R. (2006) European challenges and opportunities: The role of Europe in the internationalization of Spanish firms. *International Economic*

History Association Papers Collection (Finland: IEHC). Online at http://www.helsinki.fi/iehc2006 (accessed 13 July 2013).
Pujolar, J. (1995) Immigration in Catalonia: The politics of sociolinguistic research. *Catalan Review: International Journal of Catalan Culture* 9 (2), 141–162.
Pujolar, J. (2001) *Gender, Heteroglossia and Power: A Sociolinguistic Study of Youth Culture*. Berlin: Mouton de Gruyter.
Pujolar, J. (2007a) Bilingualism and the nation-state in the post-national era. In M. Heller (ed.) *Bilingualism: A Social Approach* (pp. 71–95). London: Palgrave Macmillan.
Pujolar, J. (2007b) African women in Catalan language courses: Struggles over class, gender and ethnicity in advanced liberalism. In B.S. McElhinny (ed.) *Words, Worlds and Material Girls: Language, Gender and Globalization* (pp. 305–348). Berlin: Mouton de Gruyter.
Pujolar, J. (2007c) The future of Catalan: Language endangerment and nationalist discourses in Catalonia. In A. Duchêne and M. Heller (eds) *Discourses of Endangerment: Interest and Ideology in the Defence of Language* (pp. 121–148). London: Continuum.
Pujolar, J. (2009) Immigrants in Catalonia: Marking territory through language. In J. Collins, S. Slembrouck and M. Baynham (eds) *Globalization and Language in Contact: Scale, Migration, and Communicative Practices* (pp. 85–105). London: Continuum.
Pujolar, J. (2010) Immigration and language education in Catalonia: Between national and social agendas. *Linguistics and Education* 21, 229–243.
Pujolar, J. (2011) Els joves i la catalanitat líquida. *L'Avenç* 373, 26–31.
Pujolar, J. and Gonzàlez, I. (2013) Linguistic 'mudes' and the de-ethnicization of language choice in Catalonia. In *Catalan in the Twenty-First Century*, special issue of *International Journal of Bilingual Education and Bilingualism* 16 (2), 138–152.
Pujolar, J., Gonzàlez i Balletbò, I., Font i Tanyà, A. and Martínez i Santmartí, R. (2010) *Llengua i joves: Usos i percepcions lingüístics de la joventut catalana*. Col·lecció Aportacions 43. Barcelona: Generalitat de Catalunya.
Qué (2009) Detenidos seis paquistaníes por financiación con fin terrorista, 21 January. Madrid: Factoría de Información.
Ràdio Hospitalet (2004) L'associació AELCOM afirma que la meitat dels locutoris de la ciutat no tenen llicència, 11 August. Online at http://www.lhdigital.cat/web/digital-h/noticia/societat/-/journal_content/56_INSTANCE_43Th/11023/360627 (accessed 7 November 2008).
Rahman, T. (1997a) The Urdu–English controversy in Pakistan. *Modern Asian Studies* 31 (1), 177–207.
Rahman, T. (1997b) Language and ethnicity in Pakistan. *Asian Survey* 37 (9), 833–839.
Rahman, T. (2002) *Language, Ideology and Power: Language-Learning Among Muslims of Pakistan and North India*. Oxford: Oxford University Press.
Ramírez, J.P. (2007) 'Aunque se fue tan lejos nos vemos todos los días': Migración transnacional y uso de nuevas tecnologías de comunicación. In C.T. Albornoz, V.J. Cabrera, K. Palacios, J.P. Ramírez and D.Q. Villafuerte (eds) *Los usos de Internet: comunicación y sociedad* (vol. 2, pp. 7–64). Quito, Ecuador: FLACSO.
Rath, J. and Kloosterman, R. (2000) Outsiders' business: A critical review of research on immigrant entrepreneurship. *International Migration Review* 34 (3), 657–681.
Rheingold, H. (2002) *Smart Mobs: The Next Social Revolution. Transforming Cultures and Communities in the Age of Instant Access*. Cambridge, MA: Perseus.
Robledo, J. (2008) Generación-i. La transición multicultural. *El País Semanal*, 6 July. Online at http://elpais.com/diario/2008/07/06/eps/1215325611_850215.html (accessed 1 July 2013).

Roca i Albert, J., Sànchez, A. and Nadal, M. (2009) *Barcelona connectada, ciutadans transnacionals: Creixements migratoris i pràctiques urbanes.* Barcelona: MUHBA and Fundació Jaume Bofill. Online at http://www.barcelonaconnectada.cat/ca (accessed 13 July 2013).

Rodríguez, N. (2008) *Educar desde el locutorio: Ayuda a que tus hijos sigan creciendo contigo.* Madrid: Plataforma.

Ros, A. and Boso, A. (2010) La población inmigrante en la nueva era digital. In M. Gimeno (ed.) *2010 e-España Informe anual sobre el desarrollo de la sociedad de la información en España* (pp. 142–148). Madrid: Fundación Orange.

Ros, A., González, E., Marín, A. and Sow, P. (2007) *Migration and Information Flows: A New Lens for the Study of Contemporary International Migration.* Working Paper Series WP07-002. Online at http://www.uoc.edu (accessed 13 July 2013).

Rovira Martínez, M., Saurí i Saula, E. and Gasull, B. (2005) *El català a les empreses: Context, pràctiques i discursos.* Barcelona: Proa.

Roy, S. (2003) Bilingualism and standardization in a Canadian call centre: Challenges for a linguistic minority community. In R. Bayley and S.R. Schecter (eds) *Language Socialization in Bilingual and Multilingual Societies* (pp. 269–285). Clevedon: Multilingual Matters.

Rozas Balbontín, P. (2003) *Gestión pública, regulación e internacionalización de las telecomunicaciones: el caso de Telefónica S.A.* Santiago de Chile: Gestión Pública ILPES.

Sabaté i Dalmau, M. (2009) Ideologies on multilingual practices at a rural Catalan school. *Sociolinguistic Studies* 3 (1), 37–60.

Sabaté i Dalmau, M. (2010) Nuevas categorizaciones sociales a través del habla: (In)migrantes en la era de la globalización en una localidad semi-rural. *Spanish in Context* 7 (2), 221–253.

Sabaté i Dalmau, M. (2012a) 'The official language of Telefónica is English': Problematising the construction of English as a lingua franca in the Spanish telecommunications sector. *Atlantis* 34 (1), 133–151.

Sabaté i Dalmau, M. (2012b) Aportaciones de la etnografía de red al estudio de un locutorio: Hacia un cambio de paradigma metodológico. In E. Codó, A. Patiño and V. Unamuno (eds) *La sociolingüística con perspectiva etnográfica en el mundo hispano: Nuevos contextos, nuevas realidades, nuevas aproximaciones,* special issue of *Spanish in Context* 9 (2), 191–218.

Sabaté i Dalmau, M. (2012c) A sociolinguistic analysis of transnational SMS practices: Non-elite multilingualism, grassroots literacy and social agency among migrant populations in Barcelona. In L.A. Cougnon and C. Fairon (eds) *SMS Communication – A Linguistic Approach,* special issue of *Lingvisticæ Investigationes* 35 (2), 318–340.

Sabaté i Dalmau, M. (2013a) Fighting exclusion from the margins: *Locutorios* as sites of social agency and resistance for migrants. In A. Duchêne, M.G. Moyer and C. Roberts (eds) *Language, Migration and Social Inequalities* (pp. 248–271). Bristol: Multilingual Matters.

Sabaté i Dalmau, M. (2013b) La gestió del multilingüisme en un espai regulat entre persones immigrades: El cas dels locutoris. In X. Vila i Moreno and E. Salvat (eds) *Noves Immigracions i Llengües* (pp. 119–146). Barcelona: MRR.

Sahul Hameed, S. (2008) The effects of mobile telephony on Singaporean society. In J.E. Katz (ed.) *Handbook of Mobile Communication Studies* (pp. 285–295). Cambridge, MA: MIT Press.

Saracho, O.N. and Spodek, B. (2008) Demythologizing the Mexican American father. *Journal of Hispanic Higher Education* 7 (2), 79–96.

Sarangi, S. and Candlin, C. (2001) 'Motivational relevancies': Some methodological reflections on social theoretical and sociolinguistic practice. In N. Coupland, S. Sarangi and C. Candlin (eds) *Sociolinguistics and Social Theory* (pp. 351–388). London: Longman.

Sarangi, S. and Roberts, C. (1999) *Talk, Work and Institutional Order: Discourse in Medical, Mediation and Management Setting*. New York: Mouton de Gruyter.

Schieffelin, B.B., Woolard, K.A. and Kroskrity, P.V. (eds) (1998) *Language Ideologies: Practice and Theory*. Oxford: Oxford University Press.

Scott, J.C. (1989) Everyday forms of resistance. *Copenhagen Journal of Asian Studies* 4, 33–62.

Secretaria de la Immigració (2009) *Perfils de les principals nacionalitats estrangeres*. Gener 2009. Departament d'Acció Social i Ciutadana. Secretaria de la Immigració. Govern de Catalunya. Online at http://www.gencat.cat/benestar/ immi/pdf/perfils/Perfils_Paisos-gener_09.pdf (accessed 13 July 2013).

Secretaria de la Immigració (2010) *La immigració en xifres. Nou cicle migratori. Juliol No. 6*. Barcelona: Generalitat de Catalunya. Departament d'Acció Social i Ciutadania. Secretaria per la Immigració. Online at http://www.gencat.cat/dasc/publica/butlletiIMMI/xifres6/la_immigracio_en_xifres_6.pdf (accessed 13 July 2013).

Serra del Pozo, P. (2008) Emprendedores inmigrantes en los barrios. *Revista de la Diputació de Barcelona Migrainfo 26*. Online at http://www.diba.cat/butlletins/detallRevista.asp?id=6616 (accessed 13 July 2013).

Serra del Pozo, P., Felip Puig, N., Pérez Arellano, A.I., Revuelto Lorda, U. and Rodríguez Alonso, L. (2003) El negocio étnico en la ciudad de Barcelona: El caso de los locutorios en el barrio de Ciutat Vella. *Proceedings IX Congreso Iberoamericano de Sistemas de Información Geográfica, Cáceres, Spain*. Online at http://www.cexeci.org (accessed 13 July 2013).

Sheller, M. (2011) Mobility. In *Sociopedia*. Madrid: International Sociological Association. Online at http://www.sagepub.net/isa/resources/pdf/Mobility.pdf (accessed 13 July 2013).

Sheller, M. and Urry, J. (2006) The new mobilities paradigm. *Environment and Planning A* 38, 207–226.

Sinatti, G. (2008) Migraciones, transnacionalismo y locus de investigación: Multi-localidad y la transición de 'sitios' a 'campos'. In C. Solé, S. Parella and L. Cavalcanti (eds) *Nuevos retos del transnacionalismo en el estudio de las migraciones* (pp. 91–112). Ministerio de Trabajo e Inmigración. Subdirección General de Información. Madrid: Grafo.

Slater, D. and Miller, D. (2000) *The Internet: An Ethnographic Approach*. Oxford: Berg.

Solé, C. (2011) Foreword. In A. Alarcón and L. Garzón (eds) *Language, Migration and Social Mobility in Catalonia* (pp. ix–xii). Leiden: Brill.

Solé, C. and Alarcón, A. (2001) *Llengua i economia a Catalunya: Anàlisi del procés de negociació de la llei 1/1998, del 7 de gener, de política lingüística per mitjà de la teoria de conjunts borrosos*. Barcelona: Institut d'Estudis Catalans.

Solé, C., Parella, S. and Cavalcanti, L. (2007) *L'empresariat immigrant a Espanya*. Col·lecció Estudis Socials 21. Barcelona: Fundació La Caixa.

Solé, C., Parella, S. and Cavalcanti, L. (eds) (2008) *Nuevos retos del transnacionalismo en el estudio de las migraciones*. Ministerio de Trabajo e Inmigración. Subdirección General de Información. Madrid: Grafo.

Solé, J., Castaño, J. and Díaz, A. (2005) Política lingüística a les empreses multinacionals i empreses de serveis públics a Catalunya. *Noves SL. Revista de Sociolingüística*. Online at http://www6.gencat.cat/llengcat/noves/hm05hivern/docs/sole.pdf (accessed 13 July 2013).

Solé Camardons, J. and Torrijos, A. (2011) Els estudis sobre empreses i llengua a Catalunya: Balanç i perspectives. In M. Strubell and I. Marí (eds) *Mercat global i Mercat local: Implicacions per al multilingüisme de l'empresa. Actes del Seminari del CUIMPB-CEL 2008* (pp. 79–111). Barcelona: Universitat Oberta de Catalunya.

Sonntag, S.K. (2005) Appropriating identity or cultivating capital? Global English in offshoring industries. *Anthropology of Work Review* 26 (1), 13–20.

Steinbock, D. (2005) *The Mobile Revolution: The Making of Mobile Services Worldwide*. London: Kogan Page.

Tagg, C. (2012) *The Discourse of Text Messaging: Analysis of SMS Communication*. London: Continuum.

Tan, P.K.W. and Rubdy, R. (2008) Introduction. In P.K.W. Tan and R. Rubdy (eds) *Language as Commodity. Global Structures, Local Marketplaces* (pp. 1–15). London: Continuum.

Taylor, P. and Bain, P. (2005) 'India calling to far away towns': The call centre labour process and globalization. *Work, Employment and Society* 19 (2), 261–282.

Thurlow, C. and Brown, A. (2003) Generation Txt? The sociolinguistics of young people's text-messaging. *Discourse Analysis Online*. Online at http://extra.shu.ac.uk/daol (accessed 13 July 2013).

Thurlow, C. and Mroczek, K. (2011) Fresh perspectives on new media sociolinguistics. In C. Thurlow and K. Mroczek (eds) *Digital Discourse: Language in the New Media* (pp. xix–xliv). Oxford: Oxford University Press.

Tilly, C. (2007) Trust networks in transnational migration. *Sociological Forum* 22 (1), 3–24.

Topix (2009) 'Defend to end!' Blog entry dated 13 February 2009. Online at http://www.topics.com/forums (accessed 10 October 2010).

Torres-Pla, J. (2012) L'evolució dels usos privats a Catalunya del 1997 al 2008. *Treballs de Sociolingüística Catalana* 22, 27–40.

Torrijos, A. (2013) Les trajectòries lingüístiques de les persones joves a Catalunya. Direcció General de Política Lingüística. Generalitat de Catalunya. Presentation at the II Jornades 'La recerca sociolingüística en l'àmbit de la llengua catalana'. Xarxa Cruscat. Barcelona: Institut d'Estudis Catalans (IEC).

Trenchs-Parera, M. and Newman, M. (2009) Diversity of language ideologies in Spanish-speaking youth of different origins in Catalonia. *Journal of Multilingual and Multicultural Development* 30 (6), 509–524.

20 Minutos (2008) Advertisement Plan Avanza, 1 December. Barcelona: Grupo 20 Minutos.

Unamuno, V. and Codó, E. (2007) Categorizar a través del habla: La construcción interactiva de la extranjeridad. *Discurso y Sociedad* 1 (1), 116–147.

Unamuno, V. and Patiño, A. (2009) 'Aquí no parlem català': Multilingualism and symbolic borderlines in urban high schools in Barcelona. Paper presented at the 3rd International Seminar on Language and Migration (AILA Migration and Language Research Network), 2–3 February, Universitat Autònoma de Barcelona, Barcelona.

UNESCO (2009) *Communication and Information: Linguistic Engineering*. United Nations Educational, Scientific and Cultural Organization. Online at http://www.unesco.org/new/en/communication-and-information (accessed 13 July 2013).

Urry, J. (2003) Social networks, travel and talk. *British Journal of Sociology* 54 (2), 155–175.

Urry, J. (2007) *Mobilities*. Cambridge: Polity Press.

Uy-Tioco, C. (2007) Overseas Filipino workers and text messaging: Reinventing transnational mothering. *Continuum Journal of Media and Cultural Studies* 21 (2), 253–265.

Ventura, F. (2006) *El pes d'un somriure. Alfabets de futur*. Manresa: Propaganda pel Fet.

Vertovec, S. (2001) Transnationalism and identity. *Journal of Ethnic and Migration Studies* (27) 4, 573–582.

Vertovec, S. (2006) *The Emergence of Super-Diversity in Great Britain*. Centre on Migration, Policy and Society Working Papers 25. Oxford: Oxford University.

Vertovec, S. (2007) Super-diversity and its implications. *Ethnic and Racial Studies* 29 (6), 1024–1054.

Vertovec, S. (2009) *Transnationalism*. London: Routledge.

Vertovec, S. (2010) Towards post-multiculturalism? Changing communities, contexts and conditions of diversity. *International Social Science Journal* 61 (199), 83–95.

Vigers, D. and Mar-Molinero, C. (2009) Spanish language ideologies in managing immigration and citizenship. In G. Extra, M. Spotti and P. Van Avermaet (eds) *Language Testing, Migration and Citizenship: Cross-National Perspectives on Integration Regimes* (pp. 167–187). London: Continuum.

Vodafone (2013) Comunidad mi país. Online at https://www.vodafone.es/particulares/es/moviles-y-fijo/tarifas/ahorro-y-control/con-quien-esta-lejos/mi-pais/comunidad-mi-pais (accessed 12 February 2013).

Waldinger, R., Aldrich H. and Ward, R. (1990) *Ethnic Entrepreneurs: Immigrant Businesses in Industrial Societies*. London: Sage.

Wee, L. (2008) Linguistic instrumentalism in Singapore. In P.K.W. Tan and R. Rubdy (eds) *Language as Commodity: Global Structures, Local Marketplaces* (pp. 31–43). London: Continuum.

Wellman, B. (2001) *The Persistence and Transformation of Community: From Neighbourhood Groups to Social Networks*. Report to the Law Commission of Canada. Wellman Associates. Online at http://homes.chass.utoronto.ca/~wellman/publications/lawcomm/lawcomm7.PDF (accessed 13 July 2013).

Wittel, A. (2000) Ethnography on the move: From field to net to internet. *Forum Qualitative Sozialforschung/Forum Qualitative Social Research* (1) 1. Online at http://www.qualitative-research.net/index.php/fqs/article/view/1131/2517#gcit (accessed 13 July 2013).

Woolard, K.A. (1989) *Double Talk: Bilingualism and the Politics of Ethnicity in Catalonia*. Stanford, CA: Stanford University Press.

Woolard, K.A. (2003) 'We don't speak Catalan because we are marginalized': Ethnic and class connotations of language in Barcelona. In R. Blot (ed.) *Language and Social Identity* (pp. 85–103). Westport, CT: Praeger.

Woolard, K.A. (2006) Language and identity choice in Catalonia: The interplay of contrasting ideologies of linguistic authority. Institute for International, Comparative and Area Studies, University of California. Online at http://www.ihc.ucsb.edu/research/identity_articles/WoolardNov5.pdf (accessed 13 July 2013).

Woolard, K.A. (2009) Linguistic consciousness among adolescents in Catalonia: A case study from the Barcelona urban area in longitudinal perspective. *Zeitschrift für Katalanistik* 22, 125–129.

Woolard, K.A. (2013) Is the personal political? Chronotypes and changing stances toward Catalan language and education. In *Catalan in the Twenty-First Century*, special issue of *International Journal of Bilingual Education and Bilingualism* 16 (2), 210–224.

Woolard, K.A. and Frekko, S.E. (2013) Catalan in the twenty-first century: Romantic publics and cosmopolitan communities. In *Catalan in the Twenty-First Century*, special issue of *International Journal of Bilingual Education and Bilingualism* 16 (2), 129–137.

Yagüe Llorente, J. (2005) *Indicadores comparados de servicios de telecomunicación (I Telefonía)*. Ministerio de Industria, Turismo y Comercio. Gobierno de España. Madrid: Subsecretaría SG Estudios.

Index

Access
 To citizenship: 30–35, 39–42
 To resources: 2, 14, 70–83
Agency (social agency): 3, 7, 9, 26, 59, 102, 172, 176
Androutsopoulos, Jannis: 2, 124, 139, 140, 142, 182

Back-stage versus front-stage: 21, 49, 67, 108, 110, 113, 114
Barcelona: 5, 8, 12, 16–17, 24, 39, 44, 45, 47, 49, 54, 61–66, 68, 74, 77, 94, 99, 107, 108, 127, 129, 132, 133, 155, 160, 178–180
Barth, Fredrik: 115, 180, 183
Basque language: 34, 46, 47
Bauman, Zygmunt: 1, 16, 42, 149, 156, 180, 183
Bilingualism: 2, 12–13, 124, 126, 128, 130–134
Blommaert, Jan: 8, 109, 140, 146, 154, 176, 183
Bourdieu, Pierre: 24, 108, 176, 184

Capital
 Linguistic: 5, 13, 26, 27, 42, 43, 109, 115, 116, 135, 137, 142, 150
 Networking: 59, 72, 97, 98, 106, 150, 172, 173
 Symbolic: 108, 115
Capitalism (late): 1, 4, 12, 25, 27, 38, 45, 109, 144, 148, 156, 170, 175
Castells, Manuel: 1, 2, 4, 7, 8, 9, 15, 30, 84, 90, 105, 148, 149, 172, 180, 185

Catalan language: 2, 6, 7, 12, 13, 18, 21, 22, 26, 27, 34, 42, 47, 52, 68, 69, 77, 108, 112–114, 116, 119–121, 124–135, 143, 174, 176
Catalonia: 2–4, 6, 7, 8–18, 24, 25, 27, 29, 30, 31, 33, 34, 38, 39, 42, 43, 48, 52, 53, 54, 59, 63, 65–67, 76, 80, 82, 86, 94, 100, 108, 113, 116, 117, 119–120, 123, 124, 125, 126, 127, 130, 131, 132, 135, 136, 148, 150, 174–178
Generalitat de Catalunya: 12, 13, 14, 17, 52, 189
Vallès Occidental: 16, 17, 20, 63–66, 68, 78, 82, 111, 123, 127–128, 161
Categorisation (social): 5, 27, 122, 124, 147, 148, 149, 150, 154, 156, 163
Clienthood: 151
Codó, Eva: xiv, xvi, 13, 69, 108, 122, 124, 130, 150, 177–178, 180, 185–186
Commodification (of language): 38, 146
Consumerism: 38, 84, 156, 172
Crisis (economic crisis; sovereign debt): 4, 30, 35, 48, 178
Critical sociolinguistics: xvi, 1, 3, 5, 24, 28, 171, 175
Customer service: 36, 46, 47, 48, 49, 50, 51, 52, 53, 62, 78, 79, 83, 110, 144, 150, 164

Data collection: xiv–xvi, 16–25
Dataveillance (see Regime)
Del Valle, José: 12, 46, 48, 121, 186–187
Digital divide: 74
Discourse: 38, 46, 121, 131, 141, 163

Duchêne, Alexandre: 3, 4, 28, 36, 38, 44, 46, 48, 79, 150, 187, 190, 195

Empowerment (see Agency)
English language: 12, 13, 14, 18, 20, 22, 34, 42, 44–45, 47, 48–51, 52, 53, 56, 58, 77, 82, 110, 113, 114, 121–124, 128, 134, 135, 138, 141, 142, 143, 144, 174
'Ethnic' business: 6, 15, 61, 67, 68, 108, 144, 148, 161, 176, 177
Europe (European Union): 5, 9, 11, 12, 14, 18, 25, 31, 32, 33, 35, 59, 152, 177
Exploitation (at the workplace): 15, 28, 62, 80, 149, 160–162, 169–170

Gal, Susan: 7, 138, 142, 146, 176, 188–189
Galician language: 34, 46, 47
Geertz, Clifford: 22, 189
Giddens, Anthony: 122, 156, 176, 180, 189
Globalisation: 1, 4, 5, 6–7, 27, 30, 45, 46, 49, 117, 146, 156, 160, 169, 171
Goffman, Erving: 5, 21, 189–190
Governmentality (see Regime)
Grassroots (alternative; bottom-up; vernacular)
  Institutions: 3, 4, 5, 14, 16, 25, 27, 32, 40, 58, 59, 60, 67, 69, 78, 80, 83, 137, 149, 150, 160, 169, 170, 173, 174
  Practices: 3, 7, 25, 27, 59, 84, 107, 108, 109, 139, 142, 144, 146, 150, 172, 175
Gumperz, John J.: 22, 142, 190

Heller, Monica: xvi, 3, 4, 24, 38, 44, 45, 46, 48, 79, 113, 150, 176, 190
Heteroglossy: 5, 27, 109, 146
Hybridity (linguistic): xiv, 27, 109, 116, 142, 146, 175

Identity (identity practice; identity politics; see also Categorisation): 5, 8, 13, 16, 26, 27, 48, 107, 122–125, 130–132, 142, 147, 148, 149, 150–159, 160–170, 175, 180
Inequality (social exclusion): 3, 7, 22, 28, 29, 59, 72, 109, 123, 160, 172, 175
Informal economy (grey market practices): 4, 14, 18, 19, 62, 73, 76, 78, 79, 80, 106, 108, 128, 153, 154, 172, 173

Information and communication technology (ICT)
  Collectivisation (practices): 25, 93–98, 173
  Internet: 1, 3, 9, 10, 25, 31, 67, 70, 74, 80, 85, 94, 96, 101, 103, 104, 105, 107, 119, 149, 157, 158, 159, 166, 168
  Mobile phone telephony: 1, 4, 9–10, 23, 24, 26, 29, 31–32, 35–38, 39, 41, 42, 45, 47, 51, 52, 53, 54, 57, 58, 73, 74, 85, 86, 88, 90, 91, 94, 96, 97, 99, 102, 110, 114, 121, 124, 144, 156, 157, 173, 177, 178, 179
  Remittances (money transfer services): 3, 4, 26, 60, 67, 74, 76, 81, 82, 106, 114, 159, 161, 166, 173
  SIM cards: 25, 26, 31, 32, 39, 40, 53, 60, 67, 72, 73, 74, 90, 91, 92, 96, 97, 106, 110, 144, 172, 173
Institutions: 1, 2, 3, 4, 5, 6, 7, 16, 25, 26, 27, 29, 30, 32, 34, 52, 59, 60–69, 72, 74, 76, 77, 78, 80, 83, 84, 108, 109, 112, 113, 116, 123, 135, 136, 144, 147, 149, 150, 159, 169, 172, 173, 174
Interculturality: 15, 25, 45, 51, 81, 82, 107, 109, 113, 117, 121, 122, 174

Jaffe, Alexandre: 19, 38, 44, 109, 142, 192

Labour force: 14–16, 19, 28, 52, 62, 79–80, 108, 114, 115, 149, 150, 153, 160, 161, 163, 170
Language economist: 46, 48, 178
Language workers: 79–80, 83, 149–150, 160, 166, 168, 170
Legitimacy: 16, 27, 99, 108, 109, 115, 116, 117, 121, 132, 142, 149, 154, 155, 164
Linguae francae: 12, 27, 46, 49, 50, 51, 108, 109, 116, 122, 136, 139, 142, 147, 152, 162
Linguistic brokers (mediators): 25, 60, 78, 79–83, 170
Linguistic diversity: 8, 12, 13, 23, 34, 38, 41, 42, 44, 45, 46, 48, 51, 54, 57, 58, 112, 113, 136, 139, 150, 174
Linguistic fetish: 34, 58, 144

Linguistic hierarchies: 5, 26, 52, 53, 56, 107, 109, 112, 128, 135, 137, 138, 142, 148, 152–154, 162, 172
Linguistic landscape: 5, 12, 24, 27, 42, 46, 47, 63, 69, 78, 109, 127, 131, 138, 144, 152, 173, 174
Linguistic (self-)discipline (regimentation): 27, 34, 109, 138, 146, 147, 175
Literacy (multiliteracies): 5, 8, 16, 27, 42, 43, 58, 78, 80, 81, 106, 107, 109, 139, 140, 142, 143, 144, 146, 148, 174, 175, 176

Marginalisation (social exclusion): 2, 3, 6, 7, 18, 27, 29, 30–34, 39, 83, 109, 122, 137, 146, 148, 160, 172, 175, 176
Martín-Rojo, Luisa: 5, 12, 13, 14, 154, 180, 194
Mass media (Spanish press): 9, 44, 69, 92, 27, 144
Migration
  Migration laws (policies): 29–34
  Migration movements (demography): 1–16
  Migration practices (communicative and otherwise): 1–16, 59–83, 84–106, 107–147, 148–175
Minority (languages): 12, 52, 53–58, 114, 121, 134–138, 174
Mobility (demographic): 3–4, 5–7, 10, 16, 18, 19, 21, 85–93, 109, 110, 129, 172
Mobile-isation (see Resistance)
Modern Standard Arabic: 12, 14, 42, 43, 47, 53, 54, 55, 56, 57, 77, 82, 110, 115, 136, 137, 140, 143, 158, 174, 180
Monolingualism (ideology and practice): 25, 27, 30, 33, 34, 41, 46, 48, 109, 112–114, 125, 132–133, 146, 174
Movistar (Telefónica): 13, 23, 29, 32, 33, 35, 36, 37, 38, 39–41, 44, 45,46, 47, 48, 50, 52, 54, 56, 57, 60, 62, 67, 68, 73, 78, 79, 92, 93, 94, 95, 98, 99, 157, 173, 178, 179, 186, 188, 189, 195, 199
Moyer, Melissa G.: 13, 14, 82, 108, 150, 187, 195

Multilingualism (ideology and practice): 3, 5, 7, 13, 14–16, 22, 27, 60, 62, 78, 79, 82–83, 176
Commercial multilingualism: 25, 30, 34, 36–38, 41–58, 82, 174
Non-elite multilingualism: 107–170, 175
Multinationals: 13, 23–26, 29, 32, 33, 34, 35, 36, 38, 39, 40–41, 45, 46, 48, 50, 51, 52, 56, 61, 62, 67–68, 73, 78, 83, 84, 85, 91, 92, 96, 98, 102, 144, 156, 160, 172, 173, 178, 179

Network ethnography: 16, 18–20
Network society: 5, 7, 8–9, 15, 16, 25, 29, 51, 78, 83, 84, 170
New economy: 4, 7, 27, 46, 49, 117, 146, 149, 156, 169, 171, 160
Numeracy: 5, 26, 42, 58, 80, 139, 144, 174, 175, 176

Orality: 109, 140
Orange: 23, 32, 35, 36, 37, 47, 50, 51, 52, 67, 68, 156

Panjabi: 12, 22, 27, 53, 108, 115, 117, 127–128, 138, 166, 174
Pennycook, Alastair: 3, 7, 24, 124, 140, 141, 146, 197
Private sector (see Telecommunications sector)
Pujolar, Joan: 2, 12, 13, 29, 48, 108, 127, 130, 131, 150, 156, 198

Racism: 2, 6, 168
Rampton, M.B.H.: 8, 146, 183
Regime (language-mediated citizenship regimes): 2, 25, 29–58, 59, 72, 83, 172, 174
Regimentation (See Linguistic self-discipline; Regime)
Repertoire (language and literacy repertoires): 8, 16, 42, 44, 53, 56, 60, 79, 107, 116, 125, 130, 133, 134, 135, 146, 148, 162, 175
Resistance: 1, 2, 3, 5, 7, 8, 16, 23, 25, 28, 58, 59, 60, 70–83, 106, 109, 116, 124, 138, 148, 150, 160, 168, 172, 173, 175

Resource (symbolic and material): 2, 19, 164, 172
Roberts, Celia: 19, 187, 200
Ros, Adela: 1, 2, 9, 10, 31, 36, 59, 63, 70, 74, 82, 176, 199
Russian: 12, 47, 53, 82, 151, 152

Sociolinguistics of globalisation: 1–7, 28, 174
Solé, Carlota: 6, 11, 14, 15, 48, 52, 61, 62, 177, 200
Spanish language: xvi, 2, 6, 7, 12, 13, 14, 16, 17, 18, 22, 23, 25, 26, 27, 28, 30, 31, 33, 34, 41, 42, 43, 44–48, 50, 51, 52, 53, 54, 56, 57, 58, 73, 78, 79, 80, 81, 82, 85, 107–147, 151, 152, 153, 154, 155, 156, 157, 160, 162, 163, 164, 166, 168, 173, 174, 176, 177
Spanish nation-state (Spanish government): 2, 3, 5, 6, 7, 11, 12, 24, 25, 27, 29–34, 35, 39, 42, 46, 48, 49, 51, 53, 58, 59, 63, 72, 73, 82, 83, 88,113, 171, 174, 178
  Ministry of the Interior: 11, 29, 31, 32, 194
Stereotype (stigma): 26, 69, 79, 125, 126, 153
Structuration (social structuration): 4, 108, 147, 168, 172, 174, 176, 180
Subversion (See Resistance)
Survival (See Resistance)

Technoliteracy: 16, 25, 79–83, 157, 177
Technopolitics: 2, 29, 30, 34, 49, 58, 83, 174
Telecommunications sector: 4, 13, 16, 23–24, 29–30, 34–58, 98–99, 171–174
Translation: 24, 25, 50, 53–58, 60, 78–80, 82, 134, 174
Translinguistic (practice): 27, 28, 108, 113, 116, 117, 122, 132, 136, 139, 142–144, 146, 162, 175
Transnationalism: 3, 5, 6–7

Urdu: 12, 22, 47, 51, 53, 78, 79, 82, 108, 110, 115, 124, 127–128, 133, 144, 157, 158, 166, 174, 180
Urry, John: 1, 8, 71, 190, 200, 201

Vertovec, Stephen: 1, 6, 8, 10, 105, 150, 156, 201–202
Vodafone: 23, 32, 35, 36, 37, 38, 40, 41, 47, 50, 52, 92–93, 94, 179, 202
Voice: 5, 33, 109, 140, 144, 152, 154, 160
  Migrant voices: 107–147

We-code: 22, 140
Woolard, Kathryn: 12, 13, 48, 127, 130, 131, 200, 202

Xenoglossy: 26, 140, 141, 142, 146

For Product Safety Concerns and Information please contact our EU Authorised Representative:

Easy Access System Europe

Mustamäe tee 50

10621 Tallinn

Estonia

gpsr.requests@easproject.com

www.ingramcontent.com/pod-product-compliance
Ingram Content Group UK Ltd.
Pitfield, Milton Keynes, MK11 3LW, UK
UKHW021941200326
4879IPUK00004B/40